Math Companion for Computer Science

Zamir Bavel

RESTON PUBLISHING COMPANY, INC.
A Prentice-Hall Company
Reston, Virginia

Library of Congress Cataloging in Publication Data

Bavel, Zamir, 1929–
 Math companion for computer science.

 Bibliography: p.
 Includes index.
 1. Mathematics — 1961–   . 2. Electronic data
processing — Mathematics. I. Title.
QA39.2.B38   510   81-22650
ISBN 0-8359-4300-3   AACR2
ISBN 0-8359-4299-6 (pbk.)

Interior design and editing/production supervision
by Ginger Sasser

© 1982 by
Reston Publishing Company, Inc.
A Prentice-Hall Company
Reston, Virginia

2189541

*Dedicated to
my father
and to the memory of my mother*

Contents

---

*These sections are somewhat more demanding in mathematical maturity, or somewhat less likely to be encountered by the computer scientist, or both.

**This section requires considerable mathematical maturity on the part of the reader. The topics here may be encountered infrequently.

---

*This section is somewhat more demanding in mathematical maturity, or somewhat less likely to be encountered by the computer scientist, or both.

**This section requires considerable mathematical maturity on the part of the reader. The topics here may be encountered infrequently.

*These sections are somewhat more demanding in mathematical maturity, or somewhat less likely to be encountered by the computer scientist, or both.

List of Symbols

(List of Symbols Continued on Next Page)

Symbol	Page	Symbol	Page
$\mathbb{P}$	60	S^*	227
$\mathbb{R}$	63	$o(G)$	232
$L \cdot C$ (set product)	64	$<$ (subgroup)	232
A^0	65	$o(a)$	234
A^n	65	$\langle a \rangle$	237
A^*	68	$\lhd$	237
A^+	69	G/N (groups)	238
$p(E)$	74	S_n	239
$\displaystyle\sum_{1-i}^{\infty}$	76	K_f	246
		$\cong$	247
$\displaystyle\sum_{1-i}^{n}$	76	$\mathcal{A}(G)$	248
		$\mathbb{P}$	264
$\dbinom{n}{r}$	91	$\mathbb{R}$	265
		$\aleph_0$	267
R^{-1} (inverse of a relation)	144	$<$ (cardinality)	272
Id (identity relation)	151	$>$ (cardinality)	272
$tc(R)$	156	$\sim$ (cardinality)	272
$\leqq$	171	$\nsim$ (cardinality)	272
$\geqq$	173	c	275
$<$	173	$\aleph$	275
$>$	173	$\forall$	281
glb	178	$\exists$	281
lub	178	$\sim$ (negation)	283
$\wedge$	178	$\neg$	283
$\vee$	178	$\equiv$	284
$f : D \rightarrow R$	184	$\nrightarrow$	285
id_A	197	$\neq$	288
χ_A	198	$\nleftrightarrow$	288
S^+	221	$\mathbb{N}$	304
λ	227	wff	320

Preface

The primary intent of this book is neither as a mathematics dictionary nor as a textbook on discrete mathematics, although to an extent it could serve in either capacity. Its main purpose is to present the mathematics needed in computer science *as needed, when encountered,* all in one volume, in an understandable, accessible and useful way.

This book is intended to be used by students of computer science, mathematics, engineering and other areas which use mathematics in a significant way, as well as by practitioners, especially those of computer science. It should be made clear, however, that the main intent of the book is not the original coverage of the needed mathematics, even though it may well be so used. The book is intended to be used more *in conjunction with* a course (past or present) or an exposition in computer science rather than *for* such a course.

The need for using this book arises mainly when a topic or subject in the *mathematics* needed for computer science is (i) not well understood, (ii) not well remembered, (iii) remembered imperfectly or with uncertainty, (iv) isolated in the reader's mind with insufficient cross-reference to other topics in mathematics and computer science, or (v) totally unknown to the reader.

The need also arises when the reader desires a survey and a summary of concepts and useful results concerning a mathematical topic. No doubt, other uses for this book are going to be found, some of them intended and others unanticipated. As was remarked in passing, this book can certainly be used as a dictionary of mathematics of sorts, although this is not one of its primary purposes. It can also be used as a textbook for a class in what is often called discrete mathematics, or mathematics preparatory to computer science. Such use requires some rearrangement of the topics and interspersing of the techniques of proof with the other topics. It also requires a careful selection of topics to be covered and, probably, provision of additional material for exercises. Nevertheless, the explanations in all the basic topics are intended to be both clear and pedagogically sound in order that this book may be used as a textbook in the classroom. Again, this is not one of the primary purposes of this book.

A main use for which this book is indeed intended is that of a self-study source. A reader putting the book to such use may cover a sequence of topics, a sequence of chapters, or the entire book.

One volume, containing every mathematical topic and treatment needed in computer science would be nice to have. It would also be impossible to have. On the one hand, space is too limited for such an enterprise. On the other, any attempt to approximate such a volume would likely result in a gigantic book difficult to use efficiently.

The criterion I used in selecting the topics was the likelihood of encounter during a computer scientific experience, whether in the field or in the classroom, and the anticipated need of associated material.

There are no doubt topics and concepts for which readers might look in this book and not find. In such a case, it would be a kindness to inform me.

The treatment of the various topics in this book is purposely uneven. Most topics, concepts and techniques are introduced slowly and gradually, with intuition-building examples, before they are presented formally, precisely, and in complete generality. In most cases, such presentations are aided with tables and figures. In particular, basic topics which often cause confusion are explained and treated as if for *initial* understanding, not assuming prior experience on the part of the reader.

Yet, some topics are presented tersely and with an eye to a survey of concepts and definitions and a summary of results. The reason is my expectation that the reader interested in such topics is likely to be ready for such a treatment and perhaps prefers it. I attempted to tailor the treatment and level of detail to the anticipated level of mathematical maturity and preparation of the reader who is likely to look up the topic or desire to study it.

Here, too, it is impossible to satisfy all. Both a mathematically mature reader and an inexperienced one might look up the same topic; the level of detail provided could not fit both.

My hope is that, in most cases, all types of readers can find their needs met and the presentation useful. In case of doubt or conflict, I often leaned toward more detail, rather than toward pithiness, but at the same time I tried to compensate by adding more advanced results and treatments. The reader need go only as far as circumstances and interest dictate.

The few topics and developments which require a more mature understanding of mathematics and/or are likely to be encountered less frequently, are starred as a warning to the reader. The larger the number of asterisks the more advanced the topic or the less frequently is it likely to be needed. (Such starred sections are not necessarily more difficult.) This warning is not intended for the reader who consults the book about a particular topic. If the topic is not starred, there is no need to take heed. On the other hand, if the topic of interest is starred, or is in a starred section, such a reader has little choice in the matter—he or she is interested in a more advanced topic or in one that is infrequently used in computer science. The warning is mostly for those who read the book, or segments thereof, sequentially; when the starred sections are reached, they may be delayed at first reading.

No topic can be covered completely in this kind of book. I had to estimate the extent to which the various topics may be needed. Here, again, there have to be some topics insufficiently covered for some readers. For that purpose, there is a list of further reading at the end of the book. In this list, more advanced treatments may be found, as well as sources for another ingredient missing in most of this book—proofs.

I have included proofs mainly when the flavor of the proof was of particular value and when the subject was that of proving theorems (Chapter 16). Otherwise, the emphasis in this book is on comprehension and usability; on informing the reader of the facts, but not on proving them.

The "why" of things is certainly attended to rather thoroughly in the basic presentations, right along with the "what" and the "how." But the formal proofs are almost always omitted, as they are not the task of this book. Those who desire proofs, not just to be convinced of the truth of the results, may find such proofs in the refernces of the list for further reading.

The order of logical precedence of topics and subtopics is not always the familiar one. In fact, there are cases in which, what was just the premise on whose basis a consequent was deduced, has become a consequent, while the former consequent has become the premise. Here, again, I was less concerned with identifying the cart and the horse than with presenting the facts and the tools in a usable way.

This book contains not only many cross-references both forwards and backwards, but also many instances of repeated definitions or brief explanations. The reason for that is the anticipated manner of use of the book. When an item is important for two separate topics, it must be included in both, because the seeker of one may never see the other. It would be disturbing to the reader who is interested in only one topic to have to find a needed definition elsewhere in the book. When the presentation of topics and chapters is not self-contained, when too many items would have to be duplicated from elsewhere in the book, the reader is given ample warning and all the references [usually between square brackets] to needed material in this book.

In each chapter, all the indented results are numbered sequentially. No distinction in numbering is made between definitions, theorems, algorithms and the like. For example, the following sequence of formal numbering appears in the text: *7.14 Algorithm, 7.15 Definition, 7.16 Theorem.*

The figures within each chapter are numbered consecutively, with no attention to any other sequence of numbers. The numbering continues across the sections in a chapter. The same is true for the tables in the chapter: they are numbered sequentially and without regard to the number of the figure or the theorem to which they are proximate.

The sections of a chapter often flow from one to the next, continuing an example or a discussion without interruption. The reason for that is a desire to alert the reader to the fact of the changed focus *in the table of contents* by a new section title. The seeker of reference may find the precise desired detail through the table of contents, without having to resort to the rather extensive index, and this can be done without damaging the continuity of exposition.

In the same vein, some definitions and results are presented in the flow of conversation, without special indentation and without an item-number. That usually happens where the usage and the context contribute to the understanding and an interruption for a formal and separate statement may damage the flow. Such instances are just as easy to recognize as the indented and numbered items, since they appear in bold-face type. Furthermore, they are given their full due in the index at the end of the book.

The order in which the chapters appear, and the order of topics within the chapters, has been dictated largely by the needs of the sequential reader. The prerequisite material dominates the sequence. There exist references to material which appears later in the book, but very few of them are needed for the basic understanding of the present topic for first reading. They are intended mainly for elaboration, comparison, edification, and cross-connections.

When a special symbol ($\in$, for example) is used for the first time, it also appears in the left margin. It is thus easy to find such initial use. The Table of Symbols lists all such instances with the page number of first occurrence.

I wish to thank Rockne Grauberger, Max Coe, Jim Merrifield, and Hal Beech for typing the manuscript, and Professor Arthur Skidmore of the University of Kansas for a critical reading of the manuscript. Professor Skidmore's efforts resulted in many helpful suggestions for improvements, distinctions to be made, and correction of errors.

I also wish to thank five more of my colleagues at the University of Kansas: Mr. Robert Boncella and Professors Tsutomu Kamimura and Jerzy Grzymala-Busse for suggesting topics to include in the book, and Professors Akira Kanda, Giora Slutzki and Jerzy Grzymala-Busse for a very meticulous proofing of the galleys.

I wish to express my deep appreciation to the people of Reston, who helped with the publication of this book. Most especially I wish to thank John Davis for discovery, enthusiasm, understanding, faith, and encouragement, and Larry Benincasa for the same, and also for his patience and forebearance.

All the people mentioned above should not be held accountable for any remaining faults, which are solely my responsibility.

Lastly, a request to the readers. An instance in which you discovered just what you were looking for, in just the right way and with just the right level of explanation, is as important for me to know about as is the case of a topic you were looking for and couldn't find, a treatment that is too detailed or not detailed enough, and any other flaw encountered. Just as valuable are notes of remaining typographical errors. (I have never known a book to have eliminated them all right away.) All such remarks will be gratefully received.

Lawrence, Kansas   Zamir Bavel

1

How To Use This Book

1.1 THE PURPOSES OF THIS COMPANION

[We use "the *Companion*" as the abbreviated title for this book.]

There are several ways in which this book may be used to an advantage, and most of them are intentional.

1. First there is the case of a mathematical term, or concept, which might have arisen in a computer science or mathematics class, in the reading of an article, in discussion with fellow workers, or in other ways. Not knowing precisely what it means may hamper further progress. The *Companion* is organized to make it easy to discover the meaning of the term.

2. Then, there is the case of needing to know a topic or subject never before encountered. The *Companion* was written with this in mind—it is geared to learning a mathematical topic for the first time.

3. The subject came up before. In fact, it may have been studied once in detail, but it is no longer fresh and is probably partly forgotten. There is a need for recalling or for brushing up.

4. What needs finding out is whether a particular fact is true. Perhaps the opposite is true, and perhaps neither. The *Companion* accommodates such an inquiry.

5. The user does indeed know something about the topic and is quite certain of it. However, it is possible that knowing a little more of the surrounding facts might be helpful in

whatever the user is doing. The *Companion* provides easy means for edification and amplification.

6. It may be helpful to know what cross-references are germane to a topic and what additional results and techniques affect it.

7. The *Companion* is also intended as a book for self-study. The user then may wish to read a chapter, a cluster of chapters, or the entire book in sequence.

8. One of the main purposes of the *Companion* is to serve as a mathematical handbook, in its various implied uses: to look up a term, a result, a connection; to find out how to perform a task; to discover the reason for things being as they are; and the like.

9. And then there is the use of the *Companion* as a textbook in the classroom. It was not intended originally to be used as such, but experience indicates that, even with the pedagogical lacks built into a multipurpose book of this sort, it renders excellent service as a textbook if it is carefully used.

The *Companion* may be useful in other ways, as well, but its primary and secondary functions have been described in the nine preceding paragraphs. There are three cautionary points which should be made here. First, most of the book is concerned with giving facts, explaining them, and presenting reasons for their structure and behavior, which might make for easy remembering. The proofs of results are not included, with some notable exceptions. The reader who wishes to see formal proofs is directed to the list for further reading at the end of the book, where appropriate references are cited.

A similar course of action should be taken by the occasional reader who finds the treatment of a topic insufficiently advanced. It is hoped that such cases are rare, but when they occur the list for further reading is a source for more advanced treatments, as well as other types of reference.

The last cautionary point to be made here concerns the starred sections. One asterisk by a section indicates either that a somewhat higher level of mathematical maturity is required from the reader of the section, or that there is a smaller likelihood that the topic will be needed in the pursuit of computer scientific experiences. A double asterisk by a section title signifies more of the same—a greater departure in either of the two directions.

1.2 FINDING THE MEANING OF THINGS

When it is desired to find the meaning of a term or the significance of a process or a result, there are two general ways to go about it.

First, if the reader has a fair idea of the context in which the query appears, the very detailed table of contents is the most likely place to start. The chapter and section titles could provide all the necessary clues.

If this route does not yield satisfactory results, or if more is desired on the subject than was yielded by the table of contents, the index (at the end of the book) should be used to identify all citations of the subject. Different type styles of index citations indicate different uses of the citations in the text. This allows the reader quicker access to the desired citation. The reader will often have to decide which of the citations in the index to pursue and which to ignore. For that purpose, it might be fruitful to read the instructions at the start of the index.

The reader should remain alert to cross-references while reading the text indicated by the table of contents or by the index, for they may lead to further illumination of related material under different names, and therefore not accessible directly from the index.

In any of the above cases, if just the meaning is sought, one should look for the term in boldface type, since this is the manner in which appear both the first mention of a term and its definition, as well as other important uses of it. (The boldface type is easy to spot on the printed page.) The meaning of the term usually appears in an intuitive explanation first, and then in a formal definition or statement.

1.3 LEARNING FOR THE FIRST TIME

When a need arises to learn a topic for the first time, a topic not known before, the user should discover first where the presentation of the topic begins. Then, it should be possible to tell from the beginning of the discourse what other concepts are needed in order to understand the presentation. The beginning of a topic can often be identified from the table of contents. If such an attempt fails, the index should be tried next.

When seeking to learn a new topic, without the context in which it normally appears, the effect is similar to that of starting in the middle. Thus, the reader risks missing part of the preparation. When an unfamiliar term, technique, or result appears in the presentation, the user should consult the index, find the locations of such an item in this book, and become familiar with it.

Such a procedure, though, may interrupt the flow of the initial reading. To avoid that, the user of the *Companion* may note such new items for a later search and, for the moment, attempt to read on to the end. This is the recommended procedure and the missing terms may be understood from the context.

It is recommended that the reader who is learning for the first time go back and read the presentation a second time, since the overview gained from the second reading often reveals additional meanings and relationships.

1.4 RECALLING, OR BRUSHING UP

When the term, subject, or technique is somewhat familiar, especially when the item in question has been mastered at one point, a brief reminder is likely to be more useful than a thorough explanation. For that purpose, the user of the *Companion* may wish to turn first to the index and scan the territory surrounding the term of interest. Often, a view of the associated references may be all that is needed.

If more is desired, the user should reference the main, or *formal,* instance of the definition, result, or technique. Once that has been located, the user can decide whether the formal result is insufficient and the surrounding material should be read. It is also possible that the user should return to the index and pursue further citations.

1.5 IS IT TRUE THAT . . . ?

When a user wishes to verify or refute an impression about a result or the meaning of a term, the task may be somewhat complicated. Where to start looking?

The reader may not remember the correct term, and then the index may be of little help, since one needs to know what to look for in order to look it up. The table of contents should then be consulted. When the title of a chapter seems a likely possibility, the list of section-titles in that chapter should be scanned, and likely sections should be located in the text and searched.

If adequate references exist and the index does lead to the desired subject matter, or if the search process otherwise succeeds in locating such coverage, the reader may first scan the formally numbered results. If the desired information still eludes, it may be necessary to read the presentation in more detail. If even that does not succeed, the reader may consult the reading list at the end of this book in pursuit of the desired knowledge.

1.6 WHAT ELSE SHOULD ONE KNOW ABOUT . . . ?

The pursuit of additional information may be done in three ways. First is the seeking of additional results on a subject partly explored. In many cases, the text provides the main results as part of the main development, and then lists a selection of useful results of a less fundamental nature. It is among the latter that the reader is likely to find the additional results desired.

Second, to find out how a concept or a process affects other topics or concepts, and how it is affected by them, the cross-references in the text and the index provide the needed clues.

Lastly, it may be desired to see if the item of interest appears in other contexts, where it may be presented from totally different points of view. As an example, the reader may wish to trace the various and quite different guises and transformations of Boolean algebra in this book. Here, again, both cross-references in the text and the index are to be used. When the index is being used for this purpose, a considerable difference between the page numbers of two citations of the same item presents a likelihood that the two citations appear in different chapters and thus may be from two different points of view. If not, at least the context of one citation is likely to be different from that of the other.

1.7 READING THE BOOK SEQUENTIALLY

When reading the entire book, or a portion of it, page-by-page, the reader may wish to postpone reading the starred sections until later. This, of course, depends on the reason for reading sequentially, but the starred sections are usually either more demanding or less likely to be helpful or needed.

It may also be prudent to ignore the cross-references the first time through, and only upon the second reading to attempt and pursue them. It is very rare that material which appears later in the book is needed earlier than it appears, because the order in which the topics appear in the book has been chosen with the sequential reader in mind. It is thus suggested that only if failure to pursue the forward reference seems to interfere with understanding should such a reference be attended to.

1.8 THE COMPANION AS A HANDBOOK

This is the chief use to which the *Companion* is likely to be put. In order to make efficient use of this book as a handbook, the user should become acquainted with its content and coverage.

It is therefore recommended that the user read the table of contents fairly closely. This may save a great deal of time and effort in looking up items in the index, an exercise which may be unnecessary because of the quicker reference in the table of contents.

1.9 THE COMPANION AS A TEXTBOOK

The instructor who finds the presentations of the more basic topics in the *Companion* attractive may wish to use this book as a textbook for such courses as "discrete mathematics," "finite mathematics," "mathematics for computer science," and the like. If so, the remarks which follow might be helpful.

First, if Chapter 16 on tools and techniques of proof is to be covered, it is recommended that the various methods of proof be interspersed with the material of the rest of the book and used on suitable theorems. Thus, both the theorem and the method of proof support each other.

Second, there are instances in which sufficient examples are provided and other instances in which the instructor is likely to need to supplement with additional examples.

Third, strong consideration should be given to skipping the starred sections completely. There is probably more material in the unstarred sections than is reasonable to present in one semester at the freshmen or sophomore level. In fact, some of the unstarred sections may also have to be abandoned.

Lastly, no exercises are provided as such, although some exercises are suggested now and then. However, the instructor may use the lists of additional results at the ends of the presentations of many topics as a pool from which to draw exercise material, especially of the type which ask students to prove such results. In addition, the instructor may construct variations on the examples given in the book as additional exercises.

It should be kept in mind, though, that the *Companion* was not written as a textbook, and therefore, if it is to be used as one, supplementary materials may be necessary. For this purpose, a Teacher's Manual is available.

2

Sets and Subsets

2.1 SETS, ELEMENTS AND MEMBERSHIP

$\in$

$\notin$

In a strict sense, sets and elements are not defined but are used as "pieces of the game." Intuitively, we may think of a **set** as a collection of objects (another undefined word) which are called **elements.** We say that an element is a **member** of (or in) the set, or that it **belongs** to the set. The symbol "$\in$" is used to denote membership in a set: if an element e belongs to a set S, we write $e \in S$. Similarly, we use "$\notin$" to denote non-membership in a set: if an element e does not belong to a set S, we write $e \notin S$. When more than one membership is to be indicated, we may use a list of elements separated by commas: $a, b, c \in S$ means $a \in S, b \in S, c \in S$.

Some examples of sets of numbers include:

a. the set of positive integers;

b. the set of nonnegative integers;

c. the set of all integers;

d. the set of integers between 4 and 7 inclusive;

e. the set of the integers 5, 172, and -3;

f. the set of rational numbers;

g. the set of real numbers;

h. the set of positive square roots of positive integers;

i. the set of prime integers;

j. the set of complex numbers with integral real part.

The following are examples of sets whose elements are not numbers:

k. the set of denominations of money (legal tender) in a country;

l. the set of planets in the solar system;

m. the set of civic clubs in town;

n. the set of the above examples from a to m.

Careful note should be taken of example **n**. First, its elements have been used as sets which themselves had elements. The same may be said about example m, since each civic club has members which may be regarded as its elements; thus a civic club is a set with regard to its members but an element with regard to the set of civic clubs in town.

2.2 DIFFICULTIES IN DESCRIBING SETS

Observe example n of the last section. We were careful not to include the set described in example n as an element in itself. If we were to allow a set to be an element of itself, we would create a paradox which would cause fatal logical difficulties beyond our present treatment. *The ability to describe a set does not guarantee its existence.* For a set to exist, it must be possible to describe it in a nonparadoxical way. It must be impossible for an object to both belong and not belong to the set. The "barber paradox" is a well known example of a condition for membership with a built-in contradiction: In a village, the barber shaves all those village men who do not shave themselves and he shaves nobody else. Who shaves the barber?

Although caution is indicated by the last paragraph, the sets we employ in this book present no such difficulty and it is safe to assume that they exist and are properly formed.

The ability to decide (by a proper description) whether an element actually does belong to a set is another matter altogether. The following two examples illustrate the point:

o. The set of theorems (provable statements) about the positive integers.

There are statements *thought* to be provable, but no proofs have been found for them. They are called **conjectures** and have been neither proved nor disproved. One such is known as *Fermat's Last Theorem:* There is no integer solution to the equation $a^n + b^n = c^n$ if $n > 2$.

p. The set of all computer programs (in a sufficiently sophisticated language, such as FORTRAN, ALGOL, PASCAL) which will reach a stop if run long enough on a computer with unrestricted memory (storage).

When such a program stops, it clearly belongs to the set. However, it is actually impossible to design an algorithm (a procedure) which can examine an arbitrary program and decide whether it will stop. (This is a version of the famous **halting problem** in the theory of computability.) Such sets are called **undecidable.**

Both sets in examples o. and p. exist as sets with nonparadoxical descriptions, even though it is not always possible to determine membership in these sets.

2.3 DESCRIBING SETS

{ }

There are a variety of ways to describe, or specify, a set in addition to such verbal descriptions employed above. One common device is the inclusion of the description in braces { } to replace the words "the set of." For instance, examples a., d., and m., above may be written, respectively, as {positive integers}, {4,5,6,7}, and {civic clubs in town}. This device is used more commonly when the elements of the set are listed, as in {4,5,6,7}. The order in which the elements are listed is not important. Repeated mention of an element does not change the membership or strengthen it; thus we may regard repetitions as deleted from the set description.

$\{x \in A: p(x)\}$

Another device is the **describing** or **defining property** which applies to some (or all) elements of a set. For instance, if I is the set of integers, we may describe the set of positive integers as $\{x \in I: x > 0\}$, which is read "the set of all x, members of the set of integers, **such that** $x > 0$." Note that the colon denotes the words "such that" and is followed by the restricting property. In the literature, "such that" is often denoted by a vertical bar, which in the above example would appear as

$$\{x \in I \,|\, x > 0\}.$$

(More on describing properties of sets may be found in Section 4.6.)

In the last example, the I formed the first restriction on the selection of elements. When such a set is clear from the context, it is not necessary to state it. Thus, we could have written instead $\{x: x \in I, x > 0\}$, reading the comma as "and." However, just $\{x: x > 0\}$ could mean, for example, all positive *real* numbers. In many cases, all elements are taken from a fixed specified set which remains as the context for the duration. Such a set is called the **universal set** (the **universe** or the

U **universe of discourse**) and is often denoted by U as a convention. Thus, if the universal set U is the set of integers, there is no ambiguity in the description $\{x: x > 0\}$ of the set of all positive integers.

2.4 THE EMPTY SET; FINITE AND INFINITE SETS

The description of a set may make it impossible for any element to belong to the set. For instance, the set $\{x \in I: x > 3, x \leq 1\}$ has no members. (Note that this set should exist, since it is possible to decide unambiguously of every potential member whether it belongs—it simply does not. This is quite different from the paradoxes mentioned earlier. Nor is the set undecidable.) No matter how it is described, the set with no elements is unique. (In other words, all sets with no elements are equal.) We call it the **empty set,** (also the **null set** or **void set**) and

$\varnothing$ denote it by $\varnothing$. The empty set plays a very important role in the manipulation of sets and should not be taken too lightly.

2.1 Theorem: The empty set is unique.

A set with only one element in it is often called a **singleton set.**

iff   A set is said to be **finite** iff (if and only if) it has a finite number of elements; otherwise it is an **infinite** set.

The cardinality of a finite set is the number of elements in the set. The cardinality of infinite sets is discussed later. We

$\#(A), |A|$ denote the cardinality of a set A both by $\#(A)$ and by $|A|$. For example, $\#(\{4,5,6,7\}) = 4$ and $\#(\varnothing) = 0$; also, $|\{4,5,6,7\}| = 4$ and $|\varnothing| = 0$.

2.5 EQUALITY OF SETS; SUBSETS AND SUPERSETS

$A = B$

$\Rightarrow$

Two sets A and B are said to be **equal,** written $A = B$, iff every element of A is also an element of B and every element of B is also an element of A; that is iff $x \in A \Rightarrow x \in B$ (read "$x \in A$ implies that $x \in B$") and $x \in B \Rightarrow x \in A$. Note the use of the biconditional "iff" indicating also that, when $x \in A \Rightarrow x \in B$ and $x \in B \Rightarrow x \in A$ then $A = B$.

> **2.2 Definition:**   Let A and B be sets. Then $A = B$ iff $(x \in A \Rightarrow x \in B)$ and $(x \in B \Rightarrow x \in A)$.

$A \neq B$

Two sets are then equal when neither can be distinguished from the other by their elements. Similarly, two sets A and B are **not equal,** written $A \neq B$, iff there is an element in one set that is not in the other.

> **2.3 Theorem:**   Let A and B be sets. Then $A \neq B$ iff there is an element x such that $(x \in A$ and $x \notin B)$ or $(x \in B$ and $x \notin A)$.

The uniqueness of the empty set, mentioned earlier in Theorem 2.1, follows from this criterion: given two empty sets, it is not possible to find an element in one which is not in the other, for neither has any elements; thus the two are equal, or "the same set."

$\subseteq, \subseteqq$

A set A is a **subset** of a set B, written $A \subseteq B$ or $A \subseteqq B$, iff every element of A is also an element of B.

> **2.4 Definition:**   Let A and B be sets. Then $A \subseteq B$ iff $x \in A \Rightarrow x \in B$.

A combination of Definitions 2.2 and 2.4 is a common **criterion for equality of sets:** Two sets A and B are equal iff $A \subseteq B$ and $B \subseteq A$. This appears as item 2 in Table 4.1, below.

$\subset$

(The reader should be cautioned that "$A \subset B$" is often used for "A is a subset of B," despite the confusion it may cause with the notation used for proper subsets (Definition 2.8, below).)

If A is a subset of B then B is a **superset** of A, written

$$B \supseteq A, \text{ or } A \subseteq B.$$

When $A \subseteq B$, we may say that A is **contained,** or is **included,** in B and that B **contains,** or **includes,** A.

When A is not a subset of B, we write $A \not\subseteq B$. By Definition 2.4, $A \not\subseteq B$ iff there exists an element of A which is not an element of B. We shall use the symbol $\exists$ to denote "there exists" or "there exist."

> **2.5 Theorem:** Let A and B be sets. Then, $A \not\subseteq B$ iff $\exists\, x \in A$ such that $x \notin B$.

Note that A and B may have common elements even when neither is a subset of the other. When A and B have no elements in common they are said to be **disjoint,** or **disjoint sets.**

In the following examples, we let $U = \{1,2,3,4,5\}$, $A = \{1,2,3\}$, $B = \{2,3,4,5\}$, $C = \{3,4\}$, $D = \varnothing$, and $E = \{4,5\}$. It should be clear that A, B, and C are subsets of U. In fact, so is D since $\varnothing \subseteq U$, as the following reasoning shows: By Theorem 2.5, to show that $\varnothing \not\subseteq U$ we must show that there exists an element of $\varnothing$ which is not an element of U. But the empty set $\varnothing$ has no elements and hence we can never find the required element. This reasoning would fit any set in place of U, and thus we have the following.

> **2.6 Theorem:** The empty set is a subset of every set.

To return to the examples, $D \subseteq A$, $D \subseteq B$, $D \subseteq C$. We also have $C \subseteq B$ but $A \not\subseteq B$, $C \not\subseteq A$, etc. Moreover, $A \subseteq A$, $B \subseteq B$, etc. by definition 2.4, since $x \in A \Rightarrow x \in A$. In fact,

> **2.7 Theorem:** Every set is its own subset and its own superset.

Thus, when $A \subseteq B$, A may equal B. To indicate that such is not the case, and that A is not empty, we say that A is a **proper subset** of B. In the above examples, C is a proper subset of B; A, B, C, and E are proper subsets of U; but D is never a proper subset, nor is B a proper subset of itself.

> **2.8 Definition:**　Let A and B be sets. Then A is a **proper subset** of B iff $A \subseteq B$, $A \neq B$, and $A \neq \varnothing$.

$\subsetneqq$

In the event A is a subset of B but not equal to it, we may write $A \subsetneqq B$, where the symbol $\neq$ indicates "not equal." Note that A may be empty and thus the symbol $\subsetneqq$ does not mean "is a proper subset."

2.6 THE POWER SET

$\mathcal{P}(A), 2^A$

On occasion, we shall be interested in a "family" of sets, rather than a single set. (The word "family" stands for "collection" or "set.") One such common case is the family of all subsets of a set A; this collection is called the **power set** of A and is denoted by $\mathcal{P}(A)$ or 2^A. As an example, where A is the set $A = \{1,2,3\}$, the power set of A is

$$\mathcal{P}(A) = \{\varnothing, \{1\}, \{2\}, \{3\}, \{1,2\}, \{1,3\}, \{2,3\}, \{1,2,3\}\}.$$

(The concept holds as well for infinite sets, but those cannot be displayed by listing.)

> **2.9 Definition:**　Let A be a set. Then, the **power set** of A is $\mathcal{P}(A) = \{B : B \subseteq A\}$.

A clue to the reason for the notation 2^A for $\mathcal{P}(A)$ may be found in the fact that the number of distinct subsets of a finite set A (i.e., the cardinality $\#(\mathcal{P}(A))$ of $\mathcal{P}(A)$) is $2^{\#(A)}$. In the example, when A has three elements, it has $2^3 = 8$ subsets. Intuitively, one might reason as follows: When we decide to construct a subset B of A, we decide for each of the elements of A whether to include it in B. Thus a subset B is created by making a binary choice ("yes" or "no") for each element of A. If A has one element, it has two subsets, one without the element, i.e. $\varnothing$, and the other with it, i.e. A. With two elements in A, say $A = \{a,b\}$, the alternatives are $\{$no,no$\} = \varnothing$, $\{$yes,no$\} = \{a\}$, $\{$no,yes$\} = \{b\}$, and $\{$yes,yes$\} = \{a,b\} = A$. In general, then, the number of subsets of A is the product of $\#(A)$ factors of 2,

$$\overbrace{2 \cdot 2 \cdot \ldots \cdot 2}^{\#(A) \text{ factors}} = 2^{\#(A)}.$$

2.10 Theorem: Where A is a finite set, $\#(\mathcal{P}(A)) = 2^{\#(A)}$.

2.7 INDEXING SETS

When references are to be made to the various members of a family of subsets, a device used often is the **indexing set.** It is a set (of the usual sort) which is used to *index* the members of a family. For instance, the set $T = \{1,2,3\}$ may be used to index three sets A_1, A_2, A_3. The set, or collection, of three sets may be written as $\{A_1,A_2,A_3\}$, or $\{A_i\}_{i \in T}$, or $\{A_i: i \in T\}$. The cardinality of the indexing set is not restricted, except that we avoid an empty indexing set.

The indexing set is often used when an arbitrary number of sets is employed, and the language may be as follows: "Let $A_i \subseteq B$, for each i in a nonempty indexing set I." This allows us to refer to subsets A_i and A_j of B without having to specify them.

The cardinality of the indexing set must fit the cardinality of the set of sets it is to index. Thus, the set $I = \{1,2,\ldots,2^{\#(A)}\}$ may be used to index the members of $\mathcal{P}(A)$, i.e. all the subsets of the finite set A. If the order of indexing is important, the correspondence between the sets and their indices must be specified as well.

3

Basic Relations and Operations on Sets

3.1 COMPLEMENTS AND SET DIFFERENCE (RELATIVE COMPLEMENT)

The **complement** of a set A is the set of the elements which are not in A. When the universal set U is obvious from the discussion, the complement of A is the set of elements of U which are not in A. In that case, U need not be specified and the notation A' or $\bar{A}$ may be used.

$A', \bar{A}$

> **3.1 Definition:** Let A be a set ($A \subseteq U$). The **complement** of A is
>
> $$A' = \{u: u \notin A\} = \{u \in U: u \notin A\}.$$

For example, if U is the set of integers and A is the set of negative integers, then A' is the set of nonnegative integers.

The **complement** of B **relative to** the set A is the set of the elements of A which are not in B. (This concept is easily misinterpreted and should be carefully studied by the beginner.) The notation used is $A - B$, and that is the reason for another name for the same concept, the **set difference.**

$A - B$

> **3.2 Definition:** Let A and B be sets. The **complement** of B **relative to** A is
>
> $$A - B = \{a \in A: a \notin B\}.$$

Note that when $A \subseteq B$, then $A - B = \varnothing$; when A and B are disjoint, $A - B = A$. Also note that $A' = U - A$.

3.2 VENN DIAGRAMS

Many relationships among sets are more easily visualized with the aid of a tool called a *Venn diagram* (after the 19th century English mathematician John Venn). The diagram uses a rectangle to represent the universal set U and closed curves, usually circles, to enclose regions which represent subsets of U. Different shadings of regions are used to indicate various subsets of U. In Figure 3.1, the complement A' of A (in U) is shaded in diagram (i) and the relative complement $A - B$ is shaded in diagram (ii). In the latter, if $A \subseteq B$ then all the elements of A are concentrated in the region common to A and B and thus the shaded region has no elements, i.e. $A - B = \varnothing$. On the other hand, if A and B are disjoint, the region common to A and B is empty and thus all members of A are concentrated in the shaded area, i.e. $A - B = A$.

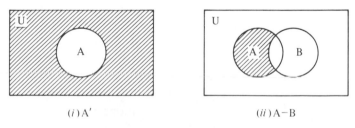

(i) A' (ii) $A - B$

Venn diagrams depicting A' and $A - B$

FIGURE 3.1

The reader should be cautioned that Venn diagrams are *intuitive* tools and do not form proofs for what may seem evident from the diagrams. One reason for that is that, without a rigorous structure, it is too easy to overlook a possibility or to mistake a situation. Another reason, partly related to the previous, is that a two-dimensional picture may not be capable of representing relations of higher dimensionality. Venn diagrams are offered here only as an *aid to visualization* of interrelationships among sets. They are not substitutes for proofs, but in this discussion they should be of value, illustrating both definitions and theorems.

3.3 THE LAW OF INVOLUTION

As an example, consider the law of involution: *the complement of the complement of a set is the set itself.*

3.3 Theorem: Let $T \subseteq S$. Then

$$S - (S - T) = T.$$

In particular,

$$(T')' = T.$$

Here we use diagrams to illustrate Theorem 3.3 in two parts. In the Venn diagrams[1] of Figure 3.2, the shaded region in diagram (i) represents $S - T$. When that is to be removed from S, as indicated in diagram (ii), all that is left is the shaded region T.

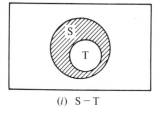 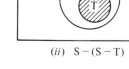

(*i*) S − T   (*ii*) S − (S − T)

Venn diagrams illustrating Theorem 3.3

FIGURE 3.2

A proof of Theorem 3.2 is presented here both to contrast it with the *illustration* by Venn diagrams and to illustrate the type of proof often used for simple relationships among sets. We shall use Definition 2.2 of equality of sets and show that any element of $S - (S - T)$ belongs to T and vice versa. (The reader may use Figure 3.2 to identify the various regions mentioned in the proof.)

Let $a \in S - (S - T)$. Then $a \in S$ and $a \notin S - T$, by Definition 3.2. But then, since $a \in S$ and $T \subseteq S$, a must be an

---

[1]These are more appropriately named "Euler diagrams" but we shall not force the distinction here.

element of T since it is not in $S - T$. Thus, $a \in S - (S - T)$ $\Rightarrow a \in T$. But then, by Definition 2.4, $S - (S - T) \subseteq T$.

Now let $a \in T$ instead. Then, since $T \subseteq S$, $a \in S$ but $a \notin S - T$, by Definition 3.2. But then, by the same definition, $a \in S - (S - T)$. Hence, $T \subseteq S - (S - T)$.

By equality of sets (Definition 2.2) and the two set inclusions, $S - (S - T) = T$. ∎

3.4 UNION AND INTERSECTION

Given one subset A of a universal set U, we were able to form one new subset A', as shown in Figure 3.1 (i). With two subsets A and B of U, we formed the difference $A - B$ shown in Figure 3.1 (ii), and we could have formed $B - A$. New subsets may be formed using the operations of **union** and **intersection** of sets.

The **union** of two sets A and B is the set of all elements of A and B; that is, if an element is in A, or in B, it is in the union of A and B. Note that an element that is both in A and in B belongs in the union for both reasons; thus we may say that an element is in the union of A and B iff it is in A or in B *or in both*.

$\cup$

3.4 Definition: Let A and B be subsets of the universal set U. Then, the **union** of A and B is

$$A \cup B = \{u \in U: u \in A \text{ or } u \in B\}.$$

The union is illustrated in Figure 3.3. Note that regions A and B are depicted with an overlapping region, since there may be elements which belong to both sets. If no such elements exist, the region is empty and A and B are said to be **disjoint**.

The elements which belong to *both* sets A and B form the **intersection** of the two sets, illustrated in Figure 3.4.

$\cap$

3.5 Definition: Let A and B be subsets of the universal set U. Then the **intersection** of A and B is

$$A \cap B = \{u \in U: u \in A \text{ and } u \in B\}.$$

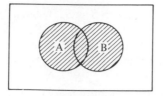

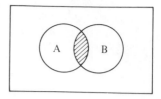

Illustrating A ∪ B

FIGURE 3.3

Illustrating A ∩ B

FIGURE 3.4

A few examples may help at this point:

a. $A = \{3,5,11\}$, $B = \{2,5,6,7\}$. Then, $A \cup B = \{2,3,5,6,7,11\}$ and $A \cap B = \{5\}$.

b. $A = \{3,5,11\}$, $B = \{3,5\}$. Then, $A \cup B = \{3,5,11\} = A$ and $A \cap B = \{3,5\} = B$.

c. $A = \{3,5,11\}$, $B = \{2,6,7\}$. Then, $A \cup B = \{2,3,5,6,7,11\}$ and $A \cap B = \varnothing$.

d. A = the set of odd integers,
   B = the set of positive integers,
   C = the set of even integers, and
   D = the set of negative integers.

Then, $A \cup B$ = the set of all positive integers and also the negative odd integers; $A \cap B$ = the set of all odd positive integers; $A \cup C$ = the set of all integers; $A \cap C = \varnothing$, the set of integers which are both odd and even; $B \cup D$ = the set of all nonzero integers; $B \cap D = \varnothing$, the set of all integers which are both positive and negative.

These examples give rise to several notes. First, we now have a convenient notation to indicate that two sets A and B are disjoint.

3.6 Theorem: Two sets A and B are disjoint iff their intersection is empty, i.e. $A \cap B = \varnothing$.

Further, the union of a set and its superset is the superset, and their intersection is the subset.

3.7 Theorem: Let $A \subseteq B$. Then,

(*i*) $A \cup B = B$;

(*ii*) $A \cap B = A$.

The Venn diagrams in Figure 3.5 illustrate the theorem.

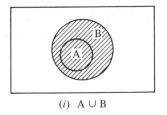

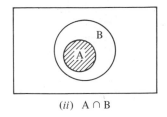

(*i*) A ∪ B   (*ii*) A ∩ B

Illustrating Theorem 3.6

FIGURE 3.5

3.5 THE SUBSETS GENERATED BY TWO
SETS

At this point, we can describe the sixteen regions of a Venn diagram with two sets. (Recall that the rectangle depicts the universal set U.) These sixteen regions depict the sixteen subsets of U which may be expressed with no more than two original subsets, A and B.

See Figure 3.6 on page 22.

There are many expressions possible for each of the sixteen regions. Particular attention is drawn to the two expressions in each of parts (vi), (viii), and (xiv); if the reader can become convinced of the equivalence of the two expressions in each case, it would be easier to remember some of the laws which are presented shortly. (A similar development with *three* subsets is left to the energetic reader.)

3.6 DISJOINTNESS AND INCLUSION

A convenient and important relationship between disjointness and inclusion of sets is the subject of Theorem 3.8 and is illustrated by Figure 3.7. (See page 23.) When the two sets A and B are disjoint, as in Figure 3.7 (i), each is a subset of the complement of the other. For example, $A \subseteq B'$, as is shown in part (ii). A similar illustration may be obtained from Figure 3.5 (ii), where $A \subseteq B$ and $A \cap B' = \varnothing$.

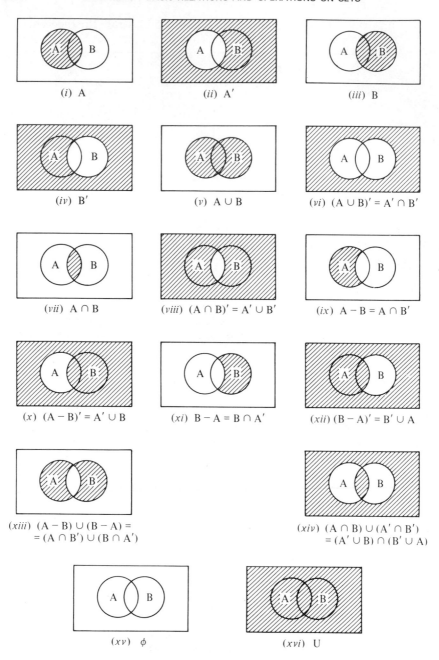

(i) A   *(ii)* A′   *(iii)* B

(iv) B′   *(v)* A ∪ B   *(vi)* (A ∪ B)′ = A′ ∩ B′

(vii) A ∩ B   *(viii)* (A ∩ B)′ = A′ ∪ B′   *(ix)* A − B = A ∩ B′

(x) (A − B)′ = A′ ∪ B   *(xi)* B − A = B ∩ A′   *(xii)* (B − A)′ = B′ ∪ A

(xiii) (A − B) ∪ (B − A) =
= (A ∩ B′) ∪ (B ∩ A′)

(xiv) (A ∩ B) ∪ (A′ ∩ B′)
= (A′ ∪ B) ∩ (B′ ∪ A)

(xv) φ   *(xvi)* U

The sixteen regions with two sets

FIGURE 3.6

3.8 Theorem: Let A and B be sets. Then,

$$A \cap B = \phi \text{ iff } A \subseteq B'.$$

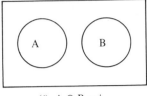 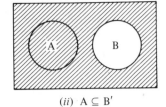

(i) $A \cap B = \phi$ (ii) $A \subseteq B'$

Illustrating Theorem 3.8

FIGURE 3.7

3.7 UNION AND INTERSECTION ARE COMMUTATIVE AND ASSOCIATIVE

A point which should be clear by now is brought to light by Theorem 3.8. The intersection operation is symmetric with regard to the two sets because the word "and" in its definition disregards the order in which the sets are listed. This property is called **commutativity** and the intersection operation is said to be commutative, as is stated in the following.

3.9 Theorem: Let A and B be sets. Then,

$$A \cap B = B \cap A.$$

As a consequence, the conclusion of Theorem 3.8 could have been stated just as well as $A \cap B = \emptyset$ iff $B \subseteq A'$.

Another set operation which is commutative is the union:

3.10 Theorem: Let A and B be sets. Then,

$$A \cup B = B \cup A.$$

To say that a binary *operation is commutative* is to say that the order of the operands does not matter. For instance, in Theorem 3.10, the union is commutative and thus the operands A and B may appear in either order with equal result. A binary operation is said to be **associative** when it does not matter which of two successive occurrences of the operation is to be performed first. Thus we have *three* operands separated by *two* occurrences of the operation, such as $A \cup B \cup C$, and the *order of performing of the operations* does not matter. In our example, performing the first union before the second results in $(A \cup B) \cup C$ while performing the second union first results in $A \cup (B \cup C)$.

Both union and intersection are associative operations on sets, as is stated in the following.

3.11 Theorem: Let A, B, and C be sets. Then,

(*i*) $(A \cup B) \cup C = A \cup (B \cup C)$;

(*ii*) $(A \cap B) \cap C = A \cap (B \cap C)$.

The associativity of the union and intersection operations, depicted in Theorem 3.11, permits the removal of the parentheses so that expressions of the forms $A \cup B \cup C$ and $A \cap B \cap C$ are not ambiguous. In fact, these results are extendable to an arbitrary (finite) number of sets, so that

$$A_1 \cup A_2 \cup \ldots \cup A_n =$$

$$\bigcup_{i=1}^{n} A_i \qquad \{u \in U: u \in A_i \text{ for some } i \in \{1, \ldots, n\}\} = \bigcup_{i=1}^{n} A_i,$$

and

$$A_1 \cap A_2 \cap \ldots \cap A_n =$$

$$\bigcap_{i=1}^{n} A_i \qquad \{u \in U: u \in A_i \text{ for every } i \in \{1, \ldots, n\}\} = \bigcap_{i=1}^{n} A_i.$$

The same holds true for the union and intersection of an arbitrary collection of sets indexed by an indexing set I. The expressions may then be written as

$$\bigcup_{i \in I} A_i, \bigcap_{i \in I} A_i \qquad\qquad \bigcup_{i \in I} A_i \text{ and } \bigcap_{i \in I} A_i.$$

With the aid of Theorem 3.11, we can similarly extend Theorems 3.9 and 3.10 to more than two sets: the order of any number of sets to be unioned (or intersected) does not matter. To illustrate with three sets, $A \cup B \cup C = (A \cup B) \cup C = (B \cup A) \cup C = B \cup (A \cup C) = B \cup (C \cup A) = (B \cup C) \cup A = (C \cup B) \cup A = C \cup B \cup A$, etc.

3.8 BOTH DISTRIBUTIVE LAWS HOLD

The meaning of **distributivity** is easy to remember from arithmetic, where multiplication is distributive over addition: The equality $2 \cdot (3 + 4) = (2 \cdot 3) + (2 \cdot 4)$ may be verified by computing on the one hand $2 \cdot (3 + 4) = 2 \cdot 7 = 14$ and on the other $(2 \cdot 3) + (2 \cdot 4) = 6 + 8 = 14$. Although this is only an example, all such examples will work since multiplication *is* distributive over addition. (The name "distributive" describes the act, or result, of distributing the multiplication onto both sides of the addition.) In contrast, addition is not distributive over multiplication, as is evident from the fact that $2 + (3 \cdot 4)$ $\neq (2 + 3) \cdot (2 + 4)$, because $2 + (3 \cdot 4) = 2 + 12 = 14$ while $(2 + 3) \cdot (2 + 4) = 5 \cdot 6 = 30$.

In the algebra of sets, each of the union and intersection operations is distributive over the other, as is stated in the following.

3.12 Theorem: Let A, B, and C be sets. Then,

(i) $A \cap (B \cup C) = (A \cap B) \cup (A \cap C)$;

(ii) $A \cup (B \cap C) = (A \cup B) \cap (A \cup C)$.

3.9 IDEMPOTENT LAWS

The **idempotent laws** in Theorem 3.13 are easy to verify from the definitions of union and intersection and to affirm with the aid of Venn diagrams. They state that a set *does not change* when unioned, or intersected, with itself.

3.13 Theorem: Let A be a set. Then,

$A \cup A = A$ and $A \cap A = A$.

3.10 LAWS OF COMPLEMENTS

Theorem 3.3 states the law of involution: for any set A, $(A')' = A$. This is one of the **laws of complements;** two more follow.

3.14 Theorem:　Let A be a set (subset of the universal set U). Then,

(*i*) $A \cup A' = U$;

(*ii*) $A \cap A' = \emptyset$.

Again, the reader may use Venn diagrams to aid the intuitive grasp of the theorem. The reader should also not neglect to reason out the statements of theorems from the definitions. For example, in Theorem 3.14, since A' has precisely those elements which are not in A, any member of U must be either in A or in A' and therefore in $A \cup A'$, implying that $U \subseteq A \cup A'$. Since all sets under consideration are subsets of U, we have the reverse inclusion $A \cup A' \subseteq U$, and therefore the equality (i) of the theorem.

In the second part of the theorem, since an element is in one of A and A' exactly when it is not in the other, no element is common to both and hence $A \cap A' = \emptyset$.

3.11 LAWS OF OPERATING WITH $\emptyset$ AND U

The **laws of operating with** $\emptyset$ **and** U have various names: **identity laws, null laws, dominance laws,** and the like. It might help the memory to realize that the empty set $\emptyset$ acts in the union as *zero* acts in addition and in the intersection as zero acts in multiplication. That is, for any number a, $a + 0 = a$ and $a \cdot 0 = 0$. Compare this to the following theorem.

3.15 Theorem:　Let A be a set. Then,

(*i*) $A \cup \emptyset = A$;

(*ii*) $A \cap \emptyset = \emptyset$.

Similarly, the universal set U acts in the intersection as 1 does in multiplication, i.e., as an *identity:* $a \cdot 1 = a$.

3.16 Theorem:　Let A be a set. Then,

$$A \cap U = A.$$

However, U does *not* act with the union as 1 acts in

addition. In fact *U dominates* the union in the way that $\varnothing$ dominates the intersection—the other set vanishes.

3.17 Theorem: Let A be a set. Then,

$$A \cup U = U.$$

It should be obvious at this point that $\varnothing$ and U are each other's complement.

3.18 Theorem:

(*i*) $\varnothing' = U$;

(*ii*) $U' = \varnothing$.

3.12 DEMORGAN'S LAWS

Two very useful tools are **DeMorgan's laws,** which were illustrated in Figure 3.6 (vi) and (viii). They state that *the complement of the union is the intersection of the complements* and that *the complement of the intersection is the union of the complements.*

3.19 Theorem: Let A and B be sets. Then,

(*i*) $(A \cup B)' = A' \cap B'$;

(*ii*) $(A \cap B)' = A' \cup B'$.

4

The Principle of Duality; Further Properties of Sets

4.1 ABSORPTION LAWS

The various laws of operating with sets, included in Chapter 3, are quite sufficient to manipulate set expressions and may be used to prove other relationships. (The summary at the end of this chapter may prove helpful both as a review and as a list of tools for set algebra.) Here, we illustrate one such important principle by providing proofs for the two **absorption laws** and then comparing the two proofs.

4.1 Theorem: Let A and B be sets. Then,

(i) $A \cup (A \cap B) = A$;

(ii) $A \cap (A \cup B) = A$.

In reasoning the results out intuitively, it becomes clear in part (i) that $A \cap B$ is part of A (it may be all of A), i.e. $A \cap B \subseteq A$, and hence the union of A and its subset $A \cap B$ is A; thus B was *absorbed*.

Similarly, A is a part of $A \cup B$, i.e. $A \subseteq A \cup B$, and hence the intersection of A and its superset $A \cup B$ is the common part, the subset A.

However, this reasoning may only show the outline of a

proof but is not itself a proof. We present the following proofs for inspection.

$$\begin{aligned}
(i)\quad A \cup (A \cap B) &= (A \cap U) \cup (A \cap B), &&\text{by } \textbf{3.16} \\
&= A \cap (U \cup B), &&\text{by } \textbf{3.12}(i) \\
&= A \cap (B \cup U), &&\text{by } \textbf{3.10} \\
&= A \cap U, &&\text{by } \textbf{3.17} \\
&= A. &&\text{by } \textbf{3.16}
\end{aligned}$$

$$\begin{aligned}
(ii)\quad A \cap (A \cup B) &= (A \cup \phi) \cap (A \cup B), &&\text{by } \textbf{3.15}(i) \\
&= A \cup (\phi \cap B), &&\text{by } \textbf{3.12}(ii) \\
&= A \cup (B \cap \phi), &&\text{by } \textbf{3.9} \\
&= A \cup \phi, &&\text{by } \textbf{3.15}(ii) \\
&= A. &&\text{by } \textbf{3.15}(i)
\end{aligned}$$

4.2 THE PRINCIPLE OF DUALITY

Apart from the fact that these proofs illustrate the use of the preceding tools, a comparison of the two proofs also suggests that one law may be obtained from the other, and one proof from the other, by

1. interchanging $\cup$ and $\cap$, and
2. interchanging ϕ and U.

In fact, applying these two interchanges to all the occurrences of $\cup$, $\cap$, ϕ, and U in the proof of (i) yields the proof of (ii).

This principle is general and is called the **principle of duality.** It uses the concept of the **dual** of a theorem, which is obtained by interchanging all occurrences of $\cup$ and $\cap$, and all occurrences of ϕ and U. Thus, for instance, the dual of $A \cap B = B \cap A$ is $A \cup B = B \cup A$; the dual of $A \cup \phi = A$ is $A \cap U = A$; and the dual of $A \cup A' = U$ is $A \cap A' = \phi$.

The principal of duality then states that, *for every proved theorem, its dual is also correct and need not be proved separately.* As a consequence, almost half of the theorems, or parts of theorems, we have detailed above need neither be listed nor memorized. Only one of a dual pair need be remembered; the principle of duality yields the other.

Caution should be exercised in applying the principle of duality to statements which make use of the subset relation

"$\subseteq$." In Theorem 3.6 it is stated that

(i) if $A \subseteq B$ then $A \cup B = B$, and

(ii) if $A \subseteq B$ then $A \cap B = A$.

In fact, the reverse implication holds in each case and the corresponding expressions are interchangeable.

4.2 Theorem: Let A and B be sets. Then,

(i) $A \subseteq B$ iff $A \cup B = B$;

(ii) $A \subseteq B$ iff $A \cap B = A$.

It should be obvious from the theorem that we would obtain false statements if we were to just interchange $\cup$ and $\cap$. There is needed also the dual of the statement $A \subseteq B$, which is not a theorem applying to all subsets A and B. In order to discover what that dual statement should be, let us examine a theorem which involves the subset relation and let us observe how the principle of duality applies there.

Consider $A \cap B \subseteq A$, which holds for all subsets A and B. This translates by Theorem 4.2 (ii) into:

$$(A \cap B) \cap A = A \cap B,$$

whose dual is:

$$(A \cup B) \cup A = A \cup B.$$

After a slight rearrangement by commutativity of "$\cup$", it becomes

$$A \cup (A \cup B) = A \cup B,$$

which translates by Theorem 4.2 (i) into

$$A \subseteq A \cup B.$$

Thus, the dual of $A \cap B \subseteq A$ should be $A \subseteq A \cup B$, or $A \cup B \supseteq A$. That is, the subset and superset relations should also be interchanged.

(The principle of duality is discussed in a more general setting in Chapter 8 in connection with partial orderings.)

4.3 SUMMARY OF SET LAWS

In what follows, an attempt is made to summarize the laws of operating with sets in a way which may be useful to the reader. Where dual pairs are obvious, they are presented together. Where the laws have names which might help recognition and memory, the names are included. A universal set U is assumed, and all sets mentioned are its subsets.

Most of the statements have been presented above and are only repeated here. However, there are several theorems which appear here for the first time, but which by now should be easy to perceive and to prove.

To make the presentation more concise and short, the following "short-hand" devices are used, each with appropriate modification of the sentence to make it fit properly.

$\forall$ "$\forall$" means "for all," "for each," "for every."

$\Rightarrow$ "$\Rightarrow$" means "imply" or "implies." This symbol is used to replace the "if" and the "then" of a statement of the form, "if statement S is true then statement T is true." It now may be written as "$S \Rightarrow T$."

$\Leftrightarrow$ "$\Leftrightarrow$" means "if and only if" or "implies, and is implied by, ..." This symbol may be used interchangeably with "iff."

[The student of logic may note that the symbol "$\Rightarrow$" and the term "implies" are used here as a predicate, not as a connective. The statement "S $\Rightarrow$ T" (which is more correctly written as " 'S' $\Rightarrow$ 'T' " or "the truth of S implies the truth of T") asserts that the truth of S does entail the truth of T. (A similar remark applies to "$\Leftrightarrow$".)]

TABLE 4.1

SUMMARY OF SET LAWS

Let A, B, C and D be arbitrary subsets of a universal set U. (Whenever these variables appear without quantification, they are to be understood to have "for all" preceding the statement in which they appear.)

Bounds: 1. $\phi \subseteq A$; $A \subseteq U$. (See 6, below.)
Equality: 2. $A = B$ iff $A \subseteq B$ and $B \subseteq A$.
Substitutivity: 3. $A = B \Rightarrow (A \cup C = B \cup C)$;
                $A = B \Rightarrow (A \cap C = B \cap C)$.

TABLE 4.1—*Continued*

Substitutivity: Continued

4. $(A = B \text{ and } C = D) \Rightarrow (A \cup C = B \cup D)$;
   $(A = B \text{ and } C = D) \Rightarrow (A \cap C = B \cap D)$.
5. $A = B \Leftrightarrow A' = B'$.

Universal bounds: 6. $\phi \subseteq A \subseteq U$. (See 1, above.)

Reflexivity of inclusion: 7. $A \subseteq A$.

Bounds with two sets: 8. $A \cap B \subseteq A \subseteq A \cup B$.

Consistency: 9. $A \subseteq B$ iff $A \cup B = B$; $A \subseteq B$ iff $A \cap B = A$.

Transitivity of inclusion: 10. $(A \subseteq B \text{ and } B \subseteq C) \Rightarrow A \subseteq C$.

Commutativity: 11. $A \cup B = B \cup A$; $A \cap B = B \cap A$.

Associativity: 12. $(A \cup B) \cup C = A \cup (B \cup C)$;
   $(A \cap B) \cap C = A \cap (B \cap C)$.

Distributivity: 13. $A \cap (B \cup C) = (A \cap B) \cup (A \cap C)$;
   $A \cup (B \cap C) = (A \cup B) \cap (A \cup C)$.

Idempotent: 14. $A \cup A = A$; $A \cap A = A$.

Involution: 15. $(A')' = A$.

Complements: 16. $A \cup A' = U$; $A \cap A' = \phi$.

Identity: 17. $A \cup \phi = A$; $A \cap U = A$.

Dominance: 18. $A \cup U = U$; $A \cap \phi = \phi$.

Complementarity of ϕ and U: 19. $\phi' = U$; $U' = \phi$.

DeMorgan's Laws: 20. $(A \cup B)' = A' \cap B'$; $(A \cap B)' = A' \cup B'$.

Absorption: 21. $A \cup (A \cap B) = A$; $A \cap (A \cup B) = A$.

22. $A \cup (A' \cap B) = A \cup B$; $A \cap (A' \cup B) = A \cap B$.

Inclusion and disjointness: 23. $A \subseteq B$ iff $A \cap B' = \phi$;
   $A \subseteq B$ iff $A' \cup B = U$.

Uniqueness of complement:

24. $(A \cup B = U \text{ and } A \cap B = \phi) \Rightarrow A = B'$.

Miscellaneous: 25. $(A \cup B = B \text{ and } A' \cap B = \phi) \Rightarrow A = B$.

26. $(A \cup B = \phi) \Rightarrow (A = \phi \text{ and } B = \phi)$;
   $(A \cap B = U) \Rightarrow (A = U \text{ and } B = U)$.

27. For any two sets A and B, if there exists a set C such that
   $A \cup C = B \cup C$ and $A \cap C = B \cap C$, then $A = B$.

28. $A \cup B = (A' \cap B')'$; $A \cap B = (A' \cup B')'$.

29. $(A \cap B) \cup (A \cap B') = A$.

30. $A \cup ((B' \cup A) \cap B)' = U$.

31. $(A \cap B) \cup (A \cap B') \cup (A' \cap B) \cup (A' \cap B') = U$.

32. $(A \cup B) \cap (A \cup B') \cap (A' \cup B) \cap (A' \cup B') = \phi$.

33. $(A \subseteq B \text{ and } A \subseteq C) \Leftrightarrow A \subseteq (B \cap C)$.

34. $(A \subseteq C \text{ and } B \subseteq C) \Leftrightarrow (A \cup B) \subseteq C$.

35. $(A \subseteq B) \Leftrightarrow (B' \subseteq A') \Leftrightarrow (A' \cup B = U) \Leftrightarrow (A \cap B' = \phi)$.

Redundancy: 36. $(A \cap B) \cup (A' \cap C) =$
   $(A \cap B) \cup (A' \cap C) \cup (B \cap C)$;
   $(A \cup B) \cap (A' \cup C) =$
   $(A \cup B) \cap (A' \cup C) \cap (B \cup C)$.

4.4 LAWS OF SET DIFFERENCE (RELATIVE COMPLEMENT)

The twelve laws which appear in Table 4.2 are the more common and useful of the laws of set difference (relative complement). (See Section 3.1, Definition 3.2.) They are all provable from Definition 3.2 and the basic laws presented in the preceding sections.

TABLE 4.2

LAWS OF SET DIFFERENCE

Let A, B and C be arbitrary subsets of the universal set U.

1. $A - B = A \cap B'$.
2. $A - \phi = A$.
3. $A - U = \phi$.
4. $A \cap (B - A) = \phi$.
5. $A \cup (B - A) = A \cup B$.
6. $A - (A \cap B) = A - B$.
7. $A - (A \cup B) = \phi$.
8. $A - B = B' - A'$.
9. $A - (B \cup C) = (A - B) \cap (A - C)$.
10. $A - (B \cap C) = (A - B) \cup (A - C)$.
11. $A \cap (B - C) = (A \cap B) - (A \cap C)$.
12. $A \cup (B - C) = (A \cup B) - (C - A)$
$= (A \cup B) - (A' \cap C)$.

4.5 THE SYMMETRIC DIFFERENCE

The **symmetric difference** of two sets A and B is sometimes called the **exclusive or** of A and B, because it is the set of elements which are in A or in B *but not in both.* For example, if $A = \{1,2,3\}$ and $B = \{3,4\}$, then the symmetric difference of A and B, written $A \oplus B$ or $A \triangle B$ is $\{1,2,4\}$. Since 3 is in both A and B, it does not belong in $A \oplus B$.

$\oplus, \triangle$

The shaded region in Figure 4.1 represents the symmetric difference of A and B.

Another way to describe $A \oplus B$ is *the union of the part of A that is not in B with the part of B that is not in A.* This is the description depicted in the following definition.

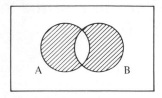

Venn diagram for A ⊕ B

FIGURE 4.1

4.3 Definition: Let A and B be two subsets of the universal set U. The **symmetric difference of A and B** is

$$A \oplus B = (A \cap B') \cup (A' \cap B).$$

Still another way to describe $A \oplus B$ is equally visible from Figure 4.1: the result of removing $A \cap B$ from $A \cup B$. As a consequence, $A \cup B$ is the result of restoring $A \cap B$ to $A \oplus B$.

4.4 Theorem: Let A and B be two subsets of the universal set U. Then,

(i) $A \oplus B = (A \cup B) - (A \cap B)$;

(ii) $A \cup B = (A \oplus B) \cup (A \cap B)$.

The following is a collection of results concerning the symmetric difference of two sets.

4.5 Theorem: Let A, B and C be subsets of the universal set U.

a. $A \oplus B = B \oplus A$.

b. $(A \oplus B) \oplus C = A \oplus (B \oplus C)$.

c. $A \oplus A = \phi$.

d. $A \oplus \phi = A$.

e. $A \oplus U = A'$.

f. $A \oplus A' = U$.

g. $(A \oplus B) \oplus B = A$.

h. $A \cap (B \oplus C) = (A \cap B) \oplus (A \cap C)$.

i. $A \cup B = A \oplus B \oplus (A \cap B)$.

j. If $A \oplus B = C$ then $B = A \oplus C$.

k. $A \oplus B = A \cup B$ iff $A \cap B = \phi$.

l. Let $A_1, A_2, \ldots, A_n, B_1, B_2, \ldots, B_n$ be subsets of U. Then

$$\left(\bigcup_{i=1}^{n} A_i \right) \oplus \left(\bigcup_{i=1}^{n} B_i \right) \subseteq \bigcup_{i=1}^{n} (A_i \oplus B_i).$$

4.5 Theorem—Continued

m. If $A \oplus B \subseteq C$ then $A \subseteq B \cup C$.

n. $(A \oplus B)' = A \oplus B' = A' \oplus B$.

o. $A' \oplus B' = (A' \oplus B)' = (A \oplus B')'$.

p. $(A \oplus B) \oplus B' = (A \oplus B') \oplus B = A'$.

q. $A \cup (B \oplus C) = (A \cup B) \oplus (A' \cap C)$.

Verification of such set-expression identities is discussed in Section 4.7, where identity **q.** is used as an example.

4.6 DESCRIBING-PROPERTIES AND OPERATIONS ON SETS; CONSTRUCTING SET EXPRESSIONS FROM COMPLICATED DESCRIPTIONS

The relationship between a set A and a "meaningful" property which describes membership in the set A (Sections 2.2 and 2.3) is governed by the **axiom of specification** of logic. The unambiguous property, "The positive integer is greater than 4," determines a set of positive integers and the property becomes the criterion for membership in the set, i.e. the set of all positive integers greater than 4. Thus, 5 is in the set because it possesses the property but 2 is not in the set because it is not greater than 4.

With a slightly enlarged scope, we have the following statement:

4.6 Theorem:
  a. (**Axiom of specification**). Every meaningful property P determines a set S_P by specifying the property P as the criterion for membership of an element in S_P.

  b. Conversely, to any set S there corresponds a property $x \in S$ of membership in S.

When set operations are used in the description of a set (such as the union of two previously identified sets), the corresponding property (condition for membership) may be expressed in a similar combination of the basic properties (membership in the first set *or* membership in the second). This technique is often useful when the description of a set is complicated.

For example, consider the description of those students who are exempt from taking the final examination in a particular course:

"All those who received more than 85 on the first test or more than 90 on the second, and in either case did not score less than 80; but if they did not score more than 95 on the homework, then they must have scored more than 90 on both first and second test."

First, we separate the description into simpler conditions:

Name of property	Name of set	Description of property
		The student scored:
a	*A*	more than 85 on test 1
b	*B*	more than 90 on test 2
c	*C*	less than 80 on test 1
d	*D*	less than 80 on test 2
e	*E*	more than 95 on the homework
f	*F*	more than 90 on test 1.

We can now construct, in a step-by-step fashion, the describing property of the set of students who are exempt from the final examination as a combination of the six simple properties. Likewise, we can express the set itself as a combination of the six corresponding sets. We match the key word "*or*" with the *union* of the corresponding sets, according to Definition 3.4. The combined condition for the first two phrases is:

"(. . . more than 85 on the first test) or (more than 90 on the second)" = "*a* or *b*."

It determines the set $A \cup B$.

" . . . did not score less than 80 on the first test" is "not *c*." It determines the set C'. Similarly, " . . . did not score less than 80 on the second test" is "not *d*." It determines the set D'.

From Definition 3.5, the "*and*" of two properties determines the *intersection* of the corresponding sets. Thus, " . . . in either case did not score less than 80" is "(not *c*) and (not *d*)" which determines the set $C' \cap D'$.

The two properties, one corresponding to "*a* or *b*" and the other to "(not *c*) and (not *d*)," are connected with an "and." Thus, the property from the beginning of the paragraph and up to the semicolon is represented by:

"(a or b) and ((not c) and (not d)),"

which determines the set $(A \cup B) \cap (C' \cap D')$.

The property, " ... did not score more than 95 on the homework," is "not e." It determines the set E'. The property, " ... they must have scored more than 90 on both first and second test," is the "and" combination "b and f," and it determines the set $B \cap F$.

The " ... if ... then ... " connection of the last two phrases of the description (following the "but") now appears as:

"if (not e) then (b and f)."

"*If p then q*" is logically equivalent to "not p or q" (see the laws of logic in Section 16.7) so that we now have:

"(not (not e)) or (b and f),"

which determines the set $(E')' \cup (B \cap F)$. The "but," immediately following the semicolon, acts as "and." Thus we have the set of students exempt from the final examination:

$$((A \cup B) \cap (C' \cap D')) \cap ((E')' \cup (B \cap F)).$$

This set expression may be simplified in the following manner. Property "b" implies property "not d," since "more than 90" automatically satisfies "not less than 80." As a result, $B \subseteq D$ and so $B \cap D' = B$. Similarly, "f" implies "a" and "a" implies "not c," and thus $F \subseteq A \subseteq C'$. This results in $A \cap C' = A$, $C' \cap F = F$ and $A \cap F = F$. These facts and the laws of sets (see summary in Section 4.3) allow the simplification:

$$((A \cup B) \cap (C' \cap D')) \cap ((E')' \cup (B \cap F)) =$$
$$((A \cup B) \cap (C' \cap D')) \cap (E \cup (B \cap F)) =$$
$$((A \cup B) \cap (C' \cap D') \cap E)$$
$$\cup ((A \cup B) \cap (C' \cap D') \cap (B \cap F)) =$$
$$(((A \cap C' \cap D') \cup (B \cap C' \cap D')) \cap E) \cup$$
$$(A \cap B \cap C' \cap D' \cap F) \cup (B \cap B \cap C' \cap D' \cap F) =$$
$$(((A \cap D') \cup (B \cap C')) \cap E) \cup (B \cap F) \cup (B \cap F) =$$
$$(A \cap D' \cap E) \cup (B \cap C' \cap E) \cup (B \cap F).$$

A helpful translation of this last set expression appears in the **modified decision table,** below. Each of the original six sets, corresponding to the elementary properties, is listed on the left. Each column of the table depicts one of the three combinations of conditions which exempt students from the final examination; each such combination signifies membership in one of the parenthesized quantities of the final expression. (These three sets may be broken into smaller subsets, which correspond to "different" combinations of conditions. However, such combinations are included in the three major ones.) The conditions in each column are simultaneous; the columns are alternatives and they are the *only* alternatives, except for "compatible" combinations (read on).

The first column corresponds to $A \cap D' \cap E$, the second to $B \cap C' \cap E$ and the third to $B \cap F$. Any other column (not showing in Table 4.3), whose entries "include" those of one of the three columns (without contradicting such implications as "*f* implies *a*") is compatible and thus also achieves exemption. All others do not.

TABLE 4.3

MODIFIED DECISION TABLE

				All compatible combinations	All other combinations
A	1				
B		1	1		
C		0			
D	0				
E	1	1			
F			1		
Exemption	1	1	1	1	0

For example, a column whose entries are 1's for A, B and F and 0 for E is compatible with the third column of the table; such a student scored more than 85 and more than 90 on test 1 and more than 90 on test 2, but not more than 95 on the homework. As another example, the combination of 1's for B and E and 0's for C and D also achieves exemption.

The correspondence between combinations of properties and combinations of sets is summarized in Table 4.4.

A similar technique is used in Section 4.7 (for a step-by-step construction of lists of regions) to verify set-expression identities.

TABLE 4.4

COMBINATIONS OF PROPERTIES AND CORRESPONDING
COMBINATIONS OF SETS

Property	Set
a	A
b	B
not a	A'
a or b	$A \cup B$
a and b	$A \cap B$
a implies b	$A \subseteq B$
if a then b	$A' \cup B$
a or b but not both	$A \oplus B$

4.7 VERIFYING SET-EXPRESSION IDENTITIES; MODIFIED VENN DIAGRAMS

A set-expression identity which involves *three* or more sets may be too difficult to verify with the common use of Venn diagrams. Consider, for example, part **q.** of Theorem 4.5 (Section 4.5):

q. $A \cup (B \oplus C) = (A \cup B) \oplus (A' \cap C)$.

The technique illustrated below uses a *modified* Venn diagram. It is based on the technique of constructing set expressions from describing properties, illustrated in Section 4.6. Each of the mutually disjoint regions is numbered, and

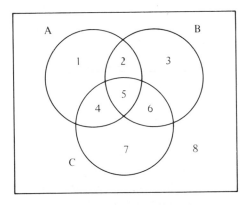

The mutually-disjoint regions of three sets

FIGURE 4.2

then these numbers may be used to build up lists which represent the set expressions. Each of the three sets may be represented by the list (set) of all the numbered regions it includes:

A: 1, 2, 4, 5.

B: 2, 3, 5, 6.

C: 4, 5, 6, 7.

Now, instead of performing operations on the sets, we perform similar operations on the corresponding lists.

To perform the *symmetric difference* on two lists of numbers, we retain from each list exactly the numbers which are in it but not in the other: $B \oplus C$: $(2,3,5,6) \oplus (4,5,6,7) =$ 2,3,4,7.

The *union* of two lists is the combined list of numbers (without repetition): $A \cup (B \oplus C)$: $(1,2,4,5) \cup (2,3,4,7) =$ $\boxed{1,2,3,4,5,7}$. This, then, is the list of regions in the expression on the left-hand-side of part **q**. of Theorem 4.5.

The right-hand-side expression is $(A \cup B) \oplus (A' \cap C)$. To construct this expression, first obtain $A \cup B$: $(1,2,4,5) \cup (2,3,5,6) = 1,2,3,4,5,6$.

The *complement* of a list is the list of missing numbers: A': $(1,2,4,5)' = 3,6,7,8$.

The *intersection* of two lists consists of the numbers in both: $A' \cap C$: $(3,6,7,8) \cap (4,5,6,7) = 6,7$.

Finally, we obtain the symmetric difference $(A \cup B) \oplus$ $(A' \cap C)$: $(1,2,3,4,5,6) \oplus (6,7) = \boxed{1,2,3,4,5,7}$.

The two lists of regions are identical, which verifies part **q**. of Theorem 4.5. (This verification process can be made rigorous enough to call it proof. This may be done with the use of the following:

Verification Theorem: Two Boolean set functions of the same *n* variables over the power set of a universal set *U* are equal iff they assume equal values for each of the 2^n assignments of ϕ's and *U*'s as values to the variables.

The importance of this theorem should not be overlooked.)

This use of lists of regions, instead of the usual illustration by Venn diagrams, is recommended for complicated cases. It may be lengthier, but the step-by-step construction of the entire expression permits great care and accuracy. However, caution must be exercised to obtain *all* regions in the diagram.

The reader should check that the number of regions is:

[2 raised to the exponent which is the number of sets].

Thus, with two sets there are $2^2 = 4$ regions, with three sets there are $2^3 = 8$ regions, and with four sets there are $2^4 = 16$ regions, as is illustrated by Figure 4.3.

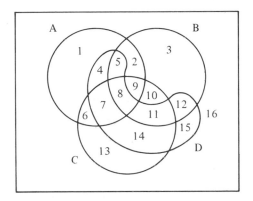

The sixteen regions of four sets

FIGURE 4.3

The rules for verification by lists of numbered regions are given in Table 4.5. Only two sets are used for the table, since no more is required to display the rules. Longer lists (for combinations of more than two sets) are obvious extensions: a single number here is replaced by the partial list of the numbered regions it represents in the larger case.

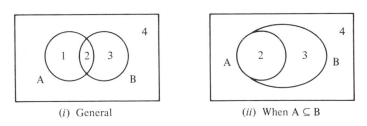

(*i*) General   (*ii*) When A ⊆ B

The four regions of two sets

FIGURE 4.4

In the following table, *a* and *b* stand for arbitrary lists of numbered regions; the operations defined on them in the table are to be regarded formally.

TABLE 4.5

MIMICKING SET OPERATIONS ON LISTS.

Set	List (general)	List (when $A \subseteq B$)
U	$u = 1,2,3,4$	$u = 2,3,4$
A	$a = 1,2$	$a = 2$
B	$b = 2,3$	$b = 2,3$
ϕ	0	0
A'	$a' = u - a = 3,4$	$a' = u - a = 3,4$

(Start with u and remove all members of a.)

$A \cup B$	$a \cup b = 1,2,3$	$a \cup b = b = 2,3$

(Every number mentioned in either or in both.)

$A \cap B$	$a \cap b = 2$	$a \cap b = 2$

(Every number mentioned in both.)

$B - A$	$b - a = (2,3) - (1,2) = 3$	$b - a = (2,3) - 2 = 3$

(Start with b and remove every number which is also in a.)

$A - B$	$a - b = (1,2) - (2,3) = 1$	$a - b = 2 - (2,3) = 0$
		(the empty list)
$A \oplus B$	$a \oplus b = (1,2) + (2,3) = 1,3$	$a \oplus b = 2 + (2,3) = 3$

(All numbers which are in either but not in both.)

U'	$u' = 0$	$u' = 0$
$A \cup \phi$	$a \cup 0 = a = 1,2$	$a \cup 0 = a = 2$
$A \cap \phi$	$a \cap 0 = 0$	$a \cap 0 = 0$
$A \oplus \phi$	$a \oplus 0 = a = 1,2$	$a \oplus 0 = a = 2$
$A - \phi$	$a - 0 = a = 1,2$	$a - 0 = a = 2$
$\phi - A$	$0 - a = 0$	$0 - a = 0$
ϕ'	$0' = u = 1,2,3,4$	$0' = u = 2,3,4$

The verification technique is illustrated once more with a four-set identity. (The numbering scheme is taken from Figure 4.3; however, once it is established, no diagram is needed.)

$$(D \cap (A - B)) \cup (A' \cap B \cap C) =$$
$$((A \cap D) \cup (B \cap C)) - (A \cap B).$$

$a = 1,2,4,5,6,7,8,9$ $b = 2,3,5,8,9,10,11,12$
$c = 6,7,8,9,10,11,13,14$ $d = 4,5,7,8,11,12,14,15$

$a - b = 1,4,6,7$

$d \cap (a - b) = (4,5,7,8,11,12,14,15) \cap (1,4,6,7) = 4,7$

$a' = u - a = 3,10,11,12,13,14,15,16$

$a' \cap b = (3,10,11,12,13,14,15,16) \cap (2,3,5,8,9,10,11,12) = 3,10,11,12$

$(a' \cap b) \cap c = (3,10,11,12) \cap (6,7,8,9,10,11,13,14) = 10,11$

$(d \cap (a - b)) \cup (a' \cap b \cap c) = (4,7) \cup (10,11) = \boxed{4,7,10,11}$.

$a \cap d = (1,2,4,5,6,7,8,9) \cap (4,5,7,8,11,12,14,15) = 4,5,7,8$

$b \cap c = (2,3,5,8,9,10,11,12) \cap (6,7,8,9,10,11,13,14) = 8,9,10,11$

$(a \cap d) \cup (b \cap c) = (4,5,7,8) \cup (8,9,10,11) = 4,5,7,8,9,10,11$

$a \cap b = (1,2,4,5,6,7,8,9) \cap (2,3,5,8,9,10,11,12) = 2,5,8,9$

$((a \cap d) \cup (b \cap c)) - (a \cap b) = (4,5,7,8,9,10,11) - (2,5,8,9) = \boxed{4,7,10,11}$.

The enterprising reader may practice on the identities of Theorem 4.5 and on the following two identities:

$$[(A \cap C) \cup (B - D)]' = (C - A) - (B \cup C),$$
$$(A \cap C - B) \cup ((B \cap D) - A) =$$
$$(A \cup B) \cap [(D - A) \cup (C - B)].$$

4.8 COUNTING ARGUMENTS FOR FINITE SETS

The term "counting arguments" applies to a variety of problems, of which we illustrate a typical example. The technique we use to address the problem is related to the one of Section 4.7.

EXAMPLE:
Of 100 students surveyed, 43 had taken course a, 55 had taken course b, 30 had taken course c, 8 had taken both courses a and b, 13 had taken both courses a and c, 15 had taken both courses b and c, and 8 had taken all three courses a, b and c. How many of the 100 students had taken none of the courses a, b and c?

We use a modified Venn diagram, with the disjoint regions generated by three sets numbered according to Figure 4.2 of Section 4.7 (reproduced here for the reader's convenience).

In the diagram of Figure 4.5, the students who had taken

course *a* are represented by the set *A*. The set consists of the mutually disjoint subsets represented by the regions 1, 2, 4 and 5. The representation is similar for course *b* and *c*, with sets *B* and *C*, respectively.

Region 5 represents $A \cap B \cap C$, the set of students who have taken all three courses. Their number is given as

$$\#(5) = \#(A \cap B \cap C) = 8.$$

There are 18 students who have taken both courses *a* and *b*. They are represented by $A \cap B$, whose region consists of both regions 2 and 5. Since these regions are disjoint, we have,

$$\#(2) + \#(5) = \#(A \cap B) = 18.$$

As a result, we have

$$\#(2) = \#(A \cap B) - \#(A \cap B \cap C) = 18 - 8 = 10.$$

(Region 2 may also be expressed as $(A \cap B) - C$.) In a similar manner we find

$$\#(4) + \#(5) = \#(A \cap C) = 13 \text{ and } \#(5) + \#(6) = \#(B \cap C) = 15.$$

Since *A* consists of the disjoint regions 1, 2, 4 and 5, we have that $\#(A) = \#(1) + \#(2) + \#(4) + \#(5)$ and hence $43 = \#(1) + 10 + 5 + 8$. We conclude that $\#(1) = 20$.

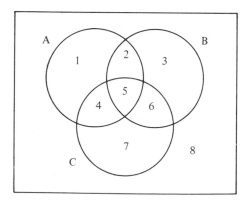

Mutually disjoint regions generated by three sets

FIGURE 4.5

In a similar manner, we arrive at $\#(3) = 30$ and $\#(7) = 10$. Therefore,

$$\#(A \cup B \cup C)$$
$$= \#(1) + \#(2) + \#(3) + \#(4) + \#(5) + \#(6) + (7)$$
$$= 20 + 10 + 30 + 5 + 8 + 7 + 10 = 90.$$

Consequently, $\#(8) = 100 - 90 = 10$.

This example is typical of many complicated problems. Another counting problem of a similar type is addressed in the next section.

4.9 THE PRINCIPLE OF INCLUSION AND EXCLUSION

Continuing with the example of Section 4.8, we wish to evaluate $\#(A \cup B)$ by first counting both $\#(A)$ and $\#(B)$. The sum is $\#(A) + \#(B) = 43 + 55 = 98$, because we counted twice the regions 2 and 5, which represent $\#(A \cap B)$. To obtain an accurate count, this number must be subtracted once, so that the intersection is counted only once:

$$\#(A \cup B) = \#(A) + \#(B) - \#(A \cap B).$$

For a similar evaluation of $\#(A \cup B \cup C)$, we first obtain $\#(A) + \#(B) + \#(C) = 43 + 55 + 30 = 128$. Regions 2 and 5 were counted both in $\#(A)$ and in $\#(B)$, so $\#(A \cap B) = 18$ should be subtracted. Similarly, $\#(A \cap C) = \#(4) + \#(5) = 13$ was counted twice, and so was $\#(B \cap C) = \#(5) + \#(6) = 15$. Subtracting the counts for the three intersections of distinct sets, we have

$$\#(A) + \#(B) + \#(C) - \#(A \cap B) - \#(A \cap C) -$$
$$\#(B \cap C) = 43 + 55 + 30 - 18 - 13 - 15 = 82.$$

Since we know that $\#(A \cup B \cup C) = 90$, we have lost 8 students. We could discover what happened if we treat the region numbers as labels and add and delete these labels from the lists of numbers according to the arithmetic operations we just performed. The original lists are

$$A = 1,2,4,5; \quad B = 2,3,5,6; \quad C = 4,5,6,7.$$

List(A) + list(B) + list(C) = 1,2,2,3,4,4,5,5,5,6,6,7. When we now delete one count each of 2 and 5 for $A \cap B$, one count each of 4 and 5 for $A \cap C$, and one count each of 5 and 6 for $B \cap C$, we get

$$\text{list}(A) + \text{list}(B) + \text{list}(C) - \text{list}(A \cap B)$$
$$- \text{list}(A \cap C) - \text{list}(B \cap C) = (1,2,2,3,4,4,5,5,5,6,6,7)$$
$$- (2,5) - (4,5) - (5,6) = 1,2,3,4, \ ,6,7.$$

The gap is where region 5 should be; it represents $A \cap B \cap C$ which was removed once too often. The correct formula must restore one count of this region:

$$\#(A \cup B \cup C) = \#(A) + \#(B) + \#(C) - \#(A \cap B)$$
$$- \#(A \cap C) - \#(B \cap C) + \#(A \cap B \cap C).$$

In the general case, the sizes of all the singleton sets are added first. Then the sizes of all intersections of *two* different sets are *subtracted*. Next, the sizes of all intersections of *three* different sets are *added,* and so on, alternating between addition and subtraction until the intersection of all the sets is either added or substracted, depending on whose turn it is.

The formula for the general case is known as the **Law of Inclusion and Exclusion:**

4.7 Theorem: Let A_i be a subset of the universal set U, for each $i \in \{1,2,\ldots,n\}$. Then,

$$\#\left(\bigcup_{i=1}^{n} A_i \right) = \sum_{i=1}^{n} \#(A_i) - \sum_{i<j} \#(A_i \cap A_j)$$
$$+ \sum_{i<j<k} \#(A_i \cap A_j \cap A_k) - \ldots + (-1)^{n-1} \#\left(\bigcap_{i=1}^{n} A_i \right).$$

The law of inclusion and exclusion, and many other counting arguments on finite sets, are based on the following principle:

The size of the union of mutually exclusive sets is the sum of the sizes of the individual sets.

This principle may be used to interpret and justify the results included in Theorem 4.8, below. They apply to many problems, including some manipulations in discrete probability, with the following translation: Use p for $\#$, *A or B* for $A \cup B$, *A and B* for $A \cap B$, *not A* for A' and *A but not B* for $A - B$.

4.8 Theorem: Let A and B be subsets of the universal set U. Then,

1. a. $\#(\phi) = 0$;
   b. $\#(A \cup B) = \#(A) + \#(B) - \#(A \cap B)$;
   c. $\#(A - B) = \#(A) - \#(A \cap B)$;
   d. $\#(B') = \#(U) - \#(B)$.
2. If $A \cap B = \phi$ (i.e., if A and B are disjoint), then

   a. $\#(A \cap B) = 0$;
   b. $\#(A \cup B) = \#(A) + \#(B)$;
   c. $\#(A - B) = \#(A)$.
3. If $B \subseteq A$, then

   a. $\#(B) \leq \#(A)$;
   b. $\#(A \cup B) = \#(A)$;
   c. $\#(A \cap B) = \#(B)$;
   d. $\#(A - B) = \#(A) - \#(B)$;
   e. $\#(B - A) = 0$.

4.10 CONCATENATION, STRINGS AND LANGUAGES

Putting letters together to make a word is an operation called **concatenation.** Digits are *concatenated* to form a telephone number, which is a **string** of digits *in order*.

When the number 2345678 is interpreted as a decimal constant, the position of each digit carries a particular value: the 7 denotes 10's, the 6 denotes 100's, etc. When the number 2345678 is interpreted as a telephone number, only the order of the digits is important. Neither fact changes the status of 2345678 as a string; in either case it is the same string of digits. A word of the English language may be regarded as a string of letters and a computer program as a string of characters. The use to which a string is put does not interfere with its status as a string.

Strings are frequent and important objects in computer science. A common use of strings occurs as input to a computer program or to a conceptual computing device. Thus, any text or data in computer memory may be regarded as a set of strings. Manipulations of data by mathematical formulas and editing the text are then *operations on strings*.

The basic elements whose concatenation forms a string are members of an agreed upon set, called the **alphabet.** The alphabet for telephone numbers is the set $\{0,1,2,3,4,5,6,7,8,9\}$, while the alphabet for words in the English language is the set of letters from a to z, both upper and lower case.

A decimal number is a **string over** the set of decimal digits, just as an English word is a string over the English alphabet.

We are concerned only with *finite* strings, which result from a finite number of concatenations of members of the alphabet. The number 2345678 is a string of **length** 7 over the set of decimal digits. The word *cat* is a string of length 3 over the English alphabet. The length of a string is the number of occurrences of members of the alphabet in the string.

A device which may appear as artificial, but which is very important to the manipulation of strings, is the **empty string.** When it stands alone, it must be denoted by a symbol, but it vanishes in concatenation with other strings. Here we denote the empty string by λ, the lower case Greek *lambda*. In the literature, it also appears as ϵ and Λ. If x is any string, then the concatenation of x with λ (on either side, and however many times λ is used) is x: $\lambda x = x\lambda = x\lambda\lambda = x$.

λ

[In the following definition, strings are defined recursively. A discussion of recursive definitions is presented in Chapter 17.]

Only finite alphabets are considered here. (Infinite alphabets are not of common interest in computer science. However, most of what is said here about finite alphabets is true for infinite ones; the exceptions are usually obvious.) Customarily, alphabets are nonempty. Our definition is slightly more general, to accomodate particular result. With these exceptions, *the reader may take an alphabet to be nonempty.*

Σ

4.9 Definition: An **alphabet** is a finite set.[1] Let Σ be an alphabet. Define a **string** (sometimes **word** or **sentence**) over Σ recursively by:

(*Basis*): λ is a string over Σ, called the **empty string,** or the **null string.** It has no elements of Σ.

(*Recursive step*): If q is a string over Σ and if $i \in \Sigma$, then the result qi of juxtaposing, or concatenating, q and i in that order is a string over Σ.

(*Disclaimer*): Only objects obtainable by a finite number of uses of the above two steps are strings over Σ.

In addition, we define the **set of all strings** over Σ and call it the **closure** of Σ (or the **Kleene closure** of Σ); it is denoted by Σ^*. ∎

Σ*

---

[1]An alphabet is usually considered nonempty.

EXAMPLES:

1. If $\Sigma = \{a,b\}$, then $\Sigma^* = \{\lambda,a,b,aa,ab,ba,bb,aaa,aab, \ldots \}$.

2. If $\Sigma = \{0,1\}$, then $\Sigma^* = \{\lambda, 0, 1, 00, 01, 10, 11, 000, 001, 010, 011, 100, \ldots \}$, which is the set of all binary strings (including the empty string).

3. If $\Sigma = \{0,1,2,3,4,5,6,7,8,9,+,-,*,/,(,)\}$, then Σ^* is harder to describe in a short space, but its members include such well-formed arithmetic expressions as $(135 + 20)*72 - (27/(16 - 7))$, as well as arithmetically unacceptable strings such as $)(/0/-172//65*-($.

4.10 Definition: Consider the set Σ^* of all strings over an alphabet Σ. The **length** of a string x over Σ is defined as the number of occurrences of elements of Σ in x: Let $x = x_1 x_2 \ldots x_n$, where $x_i \in \Sigma$, for each $i \in \{1,2, \ldots ,n\}$. The **length** of x is

$|x|$

$$|x| = |x_1 x_2 \ldots x_n| = n.$$

Additionally, the length of the empty string is $|\lambda| = 0$.

By Definition 4.9, each member of Σ is also a member of Σ^*, and thus $\Sigma \subseteq \Sigma^*$. However, i as a *member of* Σ is different from the *string i over* Σ, since they belong to different sets. In fact, in the set Σ, elements do not have length defined, while i as a member of Σ^* has length 1. Thus, in the statement $\forall i \in \Sigma, |i| = 1$, the first time i appears as a member of Σ, and the second time it appears as a member of Σ^*.

An important fact emerges from the definition of the length of a string: Σ^* *contains only strings of finite length.* This must not be confused with the fact that, *when $\Sigma \neq \phi$, Σ^* has infinitely many strings* (but each of them is of finite length!).

The following definitions and results facilitate the manipulation of arbitrary strings.

4.11 Definition: Let Σ be an alphabet and let $x,y \in \Sigma^*$, where $y = y_1 y_2 \ldots y_n$. Define the **concatenation** of x and y as the string xy obtained by repeated uses of concatenation by single elements of Σ as they appear in y; i.e.

$$
\begin{aligned}
x = x(y_1 y_2 y_3 \ldots y_n) &= (xy_1)(y_2 y_3 \ldots y_n) \\
&= ((xy_1)y_2)(y_3 \ldots y_n) \\
&= \ldots \\
&= (\ldots (((xy_1)y_2)y_3) \ldots y_n).
\end{aligned}
$$

Further, define concatenation with the empty string by:

$$\forall x \in \Sigma^*, \lambda x = x\lambda = x.$$

When concatenating several string together, only the order in which they appear is important. Thus, the parentheses in Definition 4.11 may be rearranged to yield $(\ldots(((xy_1)y_2)y_3)\ldots y_n) = x(y_1y_2y_3\ldots y_n) = xy$. This is the subject of the next theorem. (Binary operations and the associative property are discussed in detail in Section 11.3. Further discussion of concatenation and the algebraic structure of Σ^* may be found in Section 11.5.)

4.12 Theorem: Let Σ be an alphabet. Concatenation is an associative operation on Σ^*. I.e., let $x,y,z \in \Sigma^*$; then

$$(xy)z = x(yz).$$

However a string is broken into parts, the length of the string is the sum of the lengths of its parts:

4.13 Theorem: Let Σ be an alphabet and let $x,y \in \Sigma^*$. Then,

$$|xy| = |x| + |y|.$$

We have stated that there are infinitely many strings over a nonempty alphabet. This statement is formalized in Theorem 4.14, below. Here, we use the length of a string to prove this statement, as an illustration to the type of argument: Every element of Σ^* has finite length. Thus, the elements of Σ^* may be listed in a nondecreasing order of their lengths, starting with the empty string, which is the shortest one. This can be done, because there are only finitely many strings of each (finite) length.

Now, let Σ be nonempty and suppose by way of contradiction that Σ^* is finite. Then, the list of strings of Σ^* has a last member. Because of the order of listing, no other string in the list is longer than this last string on the list. Call this last string x and denote its length by $|x| = n$. We just concluded, then, that, $\forall y \in \Sigma^*, |y| \le n$. Since Σ is nonempty, it has at least one element, and we may call one such element a. But then, the concatenation xa is a string over Σ and hence xa is a member of Σ^*.

We now use the length formula of Theorem 4.13 to obtain $|xa| = |x| + |a| = n + 1$. We thus found a member of Σ^* which is longer than the longest member of Σ^*, which is a contradiction. It must be that our supposition is false; i.e., that Σ^* is indeed infinite.

This proves the following theorem.

4.14 Theorem: Σ^* is infinite if $\Sigma \neq \phi$.

When $\Sigma = \phi$, the empty string λ is still a member of Σ^*, by Definition 4.9. However, there are no elements in Σ with which to create any other string.

4.15 Theorem: If $\Sigma = \phi, \Sigma^* = \{\lambda\}$.

For fine-tuning manipulations of strings, the following definition may be helpful.

4.16 Definition: Let Σ be an alphabet and let $u = xyz$, with x, y and z arbitrary strings over Σ. Then, each of x, y, z, xy, yz is a **substring** of u. If $\lambda \neq y \neq u$, then y is a **proper substring** of u; x is a **prefix** of u, and z is a **suffix** of u. If $\lambda \neq x \neq u$, then x is a **proper prefix** of u and, if $\lambda \neq z \neq u$, then z is a **proper suffix** of u.

Note that λ is a substring of any string, as well as a prefix and suffix of every string; but λ is neither a *proper* substring, a *proper* prefix, nor a *proper* suffix of any string.

Caution should be exercised when the alphabet has elements which have the appearance of *strings* of different lengths. For example, if $\Sigma = \{1,2,12\}$, what is the length of the string 1212? It could be 2, 3, or 4, depending on its "parsing." Such ambiguity can be avoided with proper care, but the reader will be well-advised to steer clear of such cases whenever possible.

Often, some strings from Σ^* are not desirable. For instance, local telephone numbers should not start with 0 or 1; the word "qijoz" is not found in the common dictionary. Interest is more likely to exist in some particular proper subsets of Σ^*. Such sets are frequently called **languages,** in imitation of the collection of words chosen for use in a usual language.

4.17 Definition: Let Σ be an alphabet. Any subset of Σ^* is a **language** over Σ. The **empty language** is ϕ, the empty set of strings. The language whose only member is the empty string is $\Lambda = \{\lambda\}$.

Λ

The reader is cautioned not to fall in the common trap of confusing ϕ with Λ. The temptation probably originates with the two uses of the word "empty." However, the empty language (the empty set of strings) has *no* elements, while $\Lambda = \{\lambda\}$ has *one* element. In the former case, it is the *set* that is empty of strings, while in the latter, the one string in the set is the empty *string*. This distinction is highlighted by the following paraphrase of Theorem 4.15:

$$\text{If } \Sigma = \phi \text{ then } \Sigma^* = \Lambda.$$

In addition to the obvious examples of the "natural" languages and of programming languages, the following are examples of languages of some interest in the study of formal language theory. Each is over the alphabet $\{0,1\}$ and is therefore a set of **binary strings.**

1. The set of strings with an odd number of digits. (E.g. 0, 1, 011, 111. Not 01, 1101.)

2. The set of binary strings in which there are no two consecutive 0's. (E.g. λ, 0, 111, 10110. Not 100.)

3. The set of binary strings in which, if there are any 1's at all, they appear in blocks of exactly three consecutive 1's. (E.g. λ, 0, 0000, 0111001110, 111. Not 1, 011101.)

4. The set of binary strings which, when regarded as binary integers (see Chapter 14), read from left to right, are divisible by 3 (yield a remainder of 0 after division by 3). (E.g. 0, 11, 110, 1001, 1111, 10101. Not λ, 10, 100, 101, 111, 1000.)

5. The set of all strings of the form 0^n1^n, for any positive integer n.

6. The set of all strings of the form $0^n1^n0^n$, for any positive integer n.

7. The set of all binary strings of the form xx, where x is any binary string. (E.g. $\lambda = \lambda\lambda$, 00, 11, 0101, 100100.)

x^R
8. Let x^R be the **reverse** of the string x. That is, if $x = x_1x_2\ldots x_n$ then $x^R = x_n\ldots x_2x_1$ so that the reverse of 100, for instance, is 001. The example is the set of all strings of the form xx^R, where x is any binary string. (E.g. 0100110010, 0110, 00, $\lambda\lambda = \lambda$. Not 010.)

9. The set of all binary **palindromes** (strings which are identical with their reverses). I.e., the set of all binary strings x such that $x = x^R$. (E.g. Every member of Example 8 above

is a palindrome of even length. The present set contains, in addition, the palindromes of odd length: 0, 1, 010, 10101.)

The following are languages over different alphabets:

10. The set of all well-formed arithmetic expressions. (See Example 3, above.)

11. The set of all well-formed formulas of the propositional calculus. (See Section 16.7 and Examples 5 and 6 of Section 17.7.)

12. The set of addresses of living customers of a particular mail-order house.

13. The alphabet is $\Sigma = \{S, NP, VP, N, V, ADJ, ADV, ART\}$, whose symbols stand for sentence, noun-phrase, verb-phrase, noun, verb, adverb and article, respectively. $\Sigma*$ would include all constructions of types from Σ, even the unacceptable ones such as "$V\ ART\ ADV$," which could give rise to the sentence "sits the slowly." An interesting language over Σ is the set of grammatically correct constructions of types, such as "$ART\ ADJ\ N\ V\ ADV$" which, with allowable substitutions, may lead to the sentence, "The fierce lion walked majestically." (See Section 6.18.)

*4.11 INFINITE INTERSECTION AND UNION

There are areas in computer science where infinite families of sets occur frequently. Formal language theory and computability theory are two such areas. In fact, an interesting and useful characterization of $\Sigma*$, the universal set of concern in formal language theory, involves the union of an infinite family of sets. ($\Sigma*$ as the closure of an alphabet is introduced in Section 4.10; the characterization alluded to appears in Section 4.12. For *family of sets*, see Section 2.6.)

When an indexing set I (Section 2.7) is used to describe a family of sets, I could be either finite or infinite. In the event it is infinite, the family of sets indexed by members of I is likewise infinite. In that case, the notation $\underset{i \in I}{\cup} A_i$ and

---

*This section is slightly more demanding and somewhat less likely to be encountered by the computer scientist.

$\bigcap\limits_{i \in I} A_i$, used in Section 3.7, denotes the union and intersection, respectively, of an infinite family of sets.

The following definition for the union and intersection is general and includes the one for the finite case.

> **4.18 Definition:** Let I be an arbitrary nonempty indexing set and let $\{A_i: i \in I\}$ be a family of subsets of a universal set U.
>
> The set of elements $x \in U$ such that there exists at least one $i \in I$ with $x \in A_i$ is called the **union** of the family $\{A_i: i \in I\}$ and is written $\bigcup\limits_{i \in I} A_i$.
>
> The set of elements $x \in U$ such that $x \in A_i$ for every $i \in I$ is called the **intersection** of the family $\{A_i: i \in I\}$ and is written $\bigcap\limits_{i \in I} A_i$.
>
> When the indexing set I is the set of positive integers, the family of sets is a **sequence of sets;** in that case, the union is called the **countable union** and the intersection is called the **countable intersection,** and we may write[1]:

$$\bigcup\limits_{i=1}^{\infty} \quad , \quad \bigcap\limits_{i=1}^{\infty}$$

$$\bigcup\limits_{i=1}^{\infty} A_i \text{ for } \bigcup\limits_{i \in I} A_i \text{ and } \bigcap\limits_{i=1}^{\infty} A_i \text{ for } \bigcap\limits_{i \in I} A_i.$$

The binary case is the result of having only two members in the indexing set: If $I = \{1,2\}$ then $\bigcup\limits_{i \in I} A_i = A_1 \cup A_2$ and $\bigcap\limits_{i \in I} A_i = A_1 \cap A_2$.

The basic rules of operating with union and intersection are the same for the general case:

> **4.19 Theorem:** Let $\{A_i: i \in I\}$ and $\{B_j: j \in J\}$ be two families of subsets of a universal set U, where I and J are arbitrary nonempty indexing sets. Then,
>
> **a. General DeMorgan's Laws:** $\left(\bigcup\limits_{i \in I} A_i\right)' = \bigcap\limits_{i \in I} A_i'$ and $\left(\bigcap\limits_{i \in I} A_i\right)' = \bigcup\limits_{i \in I} A_i'.$

> **b. General Distributive Laws:**
>
> $$\left(\bigcup\limits_{i \in I} A_i\right) \cap \left(\bigcup\limits_{j \in J} B_j\right) = \bigcup\limits_{\substack{i \in I \\ j \in J}} (A_i \cap B_j);$$
>
> $$\left(\bigcap\limits_{i \in I} A_i\right) \cup \left(\bigcap\limits_{j \in J} B_j\right) = \bigcap\limits_{\substack{i \in I \\ j \in J}} (A_i \cup B_j).$$
>
> **c. General Associative and Commutative Laws:** Let $I = J \cup K$. Then $\bigcup\limits_{i \in I} A_i = \left(\bigcup\limits_{i \in J} A_i\right) \cup \left(\bigcup\limits_{i \in K} A_i\right)$ and $\bigcap\limits_{i \in I} A_i = \left(\bigcap\limits_{i \in J} A_i\right) \cap \left(\bigcap\limits_{i \in K} A_i\right).$

---

[1]The purist may require the union and intersection of a *sequence* of sets to be defined separately. In that case, these equivalences are theorems.

d. General Bounds Laws: Where j is any specific member of the indexing set I, $\bigcap_{i \in I} A_i \subseteq A_j \subseteq \bigcup_{i \in I} A_i$.

e. General Inclusion Laws: If $\phi \neq K \subseteq I$, i.e. K is a nonempty subset of the indexing set I, then
$$(\bigcup_{i \in K} A_i) \subseteq (\bigcup_{i \in I} A_i) \text{ and } (\bigcap_{i \in I} A_i) \subseteq (\bigcap_{i \in K} A_i).$$

f. General Difference Laws:
$$(\bigcap_{i \in I} A_i) - (\bigcup_{j \in J} B_j) = \bigcap_{i \in I} (A_i - (\bigcup_{j \in J} B_j))$$
$$= \bigcap_{i \in I} (\bigcap_{j \in J} (A_i - B_j)),$$
$$(\bigcup_{i \in I} A_i) - (\bigcup_{j \in J} B_j) = \bigcup_{i \in I} (A_i - (\bigcup_{j \in J} B_j))$$
$$= \bigcup_{i \in I} (\bigcap_{j \in J} (A_i - B_j)),$$
$$(\bigcap_{i \in I} A_i) - (\bigcap_{j \in J} B_j) = \bigcap_{i \in I} (A_i - (\bigcap_{j \in J} B_j))$$
$$= \bigcap_{i \in I} (\bigcup_{j \in J} (A_i - B_j));$$
$$(\bigcup_{i \in I} A_i) - (\bigcap_{j \in J} B_j) = \bigcup_{i \in I} (A_i - (\bigcap_{j \in J} B_j))$$
$$= \bigcup_{i \in I} (\bigcup_{j \in J} (A_i - B_j)).$$

The notation $\bigcup_{\substack{i \in I \\ j \in J}}$ in part b of Theorem 4.19 means "the union taken over all possible combinations of i from I and j from J. (With the Cartesian product $I \times J$ of Chapter 5 meaning "all possible pairs (i,j) of i from I and j from J," $\bigcup_{\substack{i \in I \\ j \in J}}$ may be written as $\bigcup_{(i,j) \in I \times J}$. A similar modification applies to the intersection.)

*4.12 CLOSURE OF FAMILIES OF SETS

The sum of two positive integers is always a positive integer. If two positive integers wish to escape from the set of positive integers, addition is not the way to do it. The set of positive integers is **closed under** addition. However, since $5 - 8 = -3$ is not a positive integer, while 5 and 8 are, the set of positive integers is *not closed* under subtraction; this is a case of two members of the set producing a nonmember by subtraction and thus escaping the original set. (Of course, there are infinitely many such examples.)

The set of all integers is, indeed, closed under subtraction

---
*This section is slightly more demanding in mathematical maturity.

(and addition, of course), but not under division. For example, 5/3 is not an integer and 1/0 is not defined at all.

The property of a set being closed under an operation (relation, function) is called **closure,** or a **closure property.** It is a very important property which participates in the definition of many concepts and structures. (Closure in an algebraic setting is discussed in Section 11.1.)

The concept of **closure of a family of sets** under a set operation is not different in principle, but it deserves special mention. Consider the family $\mathcal{F}_1 = \{\{1,2,3\}, \{3,4\}, \{1,2,3,4\}\}$ of subsets of the universal set $U = \{1,2,3,4\}$. No matter which two sets are unioned, the result is a member of $\mathcal{F}_1$: $\{1,2,3\} \cup \{1,2,3\} = \{1,2,3\}$, $\{1,2,3\} \cup \{3,4\} = \{1,2,3,4\}$, etc. Thus, $\mathcal{F}_1$ is closed under unions. On the other hand, $\{1,2,3\} \cap \{3,4\} = \{3\} \notin \mathcal{F}_1$, and consequently $\mathcal{F}_1$ is not closed under intersections.

As another example, consider the family $\mathcal{F}_2 = \{\phi,\{1,2\},\{3,4\},\{1,2,3,4\}\}$ of subsets of U. Here, not only is the union of any two members of $\mathcal{F}_2$ itself a member, but so is the *intersection:* $\{1,2\} \cap \{3,4\} = \phi \in \mathcal{F}_2$, $\{1,2\} \cap \{1,2,3,4\} = \{1,2\} \in \mathcal{F}_2$, etc. Moreover, even the *set difference* cannot aid in the escape from $\mathcal{F}_2$: $\{1,2,3,4\} - \{1,2\} = \{3,4\} \in \mathcal{F}_2$, $\{1,2\} - \{1,2,3,4\} = \phi \in \mathcal{F}_2$, etc. Thus, $\mathcal{F}_2$ is also closed under set differences, i.e. relative complements.

Lastly, the complement of any member of $\mathcal{F}_2$ is itself a member of $\mathcal{F}_2$: $\phi' = U$, $\{1,2\}' = \{3,4\}$, etc. Therefore, $\mathcal{F}_2$ is also closed under *complementation*. (The union, intersection and set difference are called binary operations on sets, since each requires *two* sets to produce a result. Complementation is a unary operation, since it is defined on *one* set at a time.)

4.20 Definition[1]: Let $\mathcal{F}$ be a family of subsets of a universal set U, let # be a binary operation[2] defined on the members of $\mathcal{F}$ and let % be a unary operation[3] defined on the

---

[1]In addition, closure is defined for families of sets under relations and under functions (mappings) to structures other than that same family, or even to structures which are not families of sets altogether. Likewise, closure is defined for families of objects other than sets. The concept is similar and is a natural extension of the present one.

[2]The notation $A \# B$, used here, is common for binary operations and is used for the union, intersection and set difference, among others. However, there are some advantages to the "functional notation" $\#(A,B)$, which is less common for a binary operation.

[3]For a unary operation, we use here the functional notation $\%(A)$. (See Chapter 9.) Yet, the customary operation symbol for the unary operation of complementation appears either after the argument (A') or above it ($\overline{A}$). Other unary operations have other customary locations for their operation symbols.

4.20 Definition—Continued
members of $\mathcal{F}$. $\mathcal{F}$ is **closed under** # (alt., **closed with respect to** #) iff $A \# B \in \mathcal{F}, \forall A, B \in \mathcal{F} \cdot \mathcal{F}$ is **closed under** % (alt., **closed with respect to** %) iff $\%(A) \in \mathcal{F}, \forall A \in \mathcal{F}$.

The concept of the **simultaneous closure** of a set under several operations is a natural extension of the (simple) closure. It is more general than its presentation in the following definition, but the latter is all that is likely to be useful to the computer scientist.

4.21 Definition: Let $\mathcal{F}$ be a family of subsets of a universal set, let $\#_1, \#_2, \ldots, \#_m$ be binary set operations and let $\%_1, \%_2, \ldots, \%_n$ be unary set operations, for some positive integers m and n. Then, $\mathcal{F}$ is **(simultaneously)closed under** $\#_1, \#_2, \ldots, \#_m$ and $\%_1, \%_2, \ldots, \%_n$ iff,
   $\forall A, B \in \mathcal{F}, A \#_i B \in \mathcal{F}$ for every $i \in \{1,2,...,m\}$, and
   $\forall A \in \mathcal{F}, \%_i(A) \in \mathcal{F}$ for every $i \in \{1,2,...,n\}$.

Examples of simultaneous closure appear in the next section.

*4.13 DEFINING SETS BY CLOSURE

Consider the *set closure,* or *Kleene closure,* of an alphabet. (See Definition 4.9 of Section 4.10.) For the sake of illustration, let $\Sigma = \{0,1\}$ and thus Σ^* is the set of all *binary strings*. The reason for the word "closure" in the name for Σ^* is not just that the set Σ^* is closed under the binary operation of concatenation of elements of Σ^*, although that is certainly true since the concatenation of any two binary strings is itself a binary string. The reason Σ^* is called the *set closure* of Σ is different: Regard the set Σ^* as one we shall build up, starting with just the empty string λ. The operation we use is **concatenation on the right with 0 or 1 (cw0o1,** for short). It is defined not only on the one object, λ, we already have in our collection, but also *on all the objects which can be produced* with this concatenation from our starting point!

cw0o1

Applying cw0o1 on the member of $\{\lambda\}$, we get $\{\lambda,0,1\}$ in the first stage. Using cw0o1 on the result of the first stage, we

---

*This section is somewhat more demanding in mathematical maturity.

obtain the additional strings 00, 01, 10, 11 and thus we enlarge our collection to $\{\lambda,0,1,00,01,10,11\}$. Obviously, there is no end to this process, since concatenation on each stage produces new, and longer, strings for the next stage. Thus, none of the increasingly larger finite collections of strings, which result after a finite number of stages, is closed under cw0o1. However, Σ^* itself is certainly closed under cw0o1, since the result of such concatenation of any binary string with 0 or with 1 is a binary string. In fact, Σ^* *is the smallest set which contains* λ *and which is closed under* cw0o1.

(Another way to describe Σ^*, which may be a closer fit to the name "the closure of Σ" is: The smallest set which includes (as strings) λ and the members of Σ and which is closed under arbitary finite concatenation.")

This is what the **closure of a set** (under an operation) means. It may be used to define a set from an initial generating subset and an operation. Given a set S and an operation $\#$ on S, one can refer to the **closure of S under** $\#$, even if S is not already closed under $\#$. (We shall use the nonstandard notation $(S,\#)^*$ for the closure of S under $\#$.) If S is already closed under $\#$, then $(S,\#)^* = S$. Otherwise, a larger set results, which is a superset of S.

Thus, for example, the closure $(\mathbb{P},-)^*$ of the set of positive integers under subtraction is the set of all integers. Likewise, the closure $(\text{ODDP},+)^*$ of the set ODDP of odd positive integers under addition is the set $\mathbb{P}$ of all positive integers. On the other hand, the closure $(\text{EVENP},+)^*$ of the set EVENP of even positive integers under addition is just the set EVENP itself.

A similar process may be used to define a family of sets, using closure: Start with a basic family of sets, called a generating family, and with a set operation (or several operations). Then "repeatedly" include any new members which can be obtained by the operation(s) on the existing members of the family. Imagine this process continuing forever, or until no new members can be added.

In the case of a finite universal set, the process ends and the closure is obtained after a finite number of steps. For example, consider the family

$$\mathcal{G} = \{\{1,2,3\},\{2,4\}\} \text{ of subsets of } U = \{1,2,3,4\}.$$

To close $\mathcal{G}$ under finite unions, we add $\{1,2,3,4\}$. The closure of $\mathcal{G}$ under *finite unions* may be written, using out nonstandard notation, as

$$(\mathcal{G}, \cup)^* = \{\{1,2,3\},\{2,4\},\{1,2,3,4\}\}.$$

Similarly, the closure of $\mathcal{G}$ under *finite intersections* is

$$(\mathcal{G}, \cap)^* = \{\{1,2,3\},\{2,4\},\{2\}\}.$$

The (simultaneous) closure of $\mathcal{G}$ under *both finite unions and finite intersections* is

$$(\mathcal{G}, \cup, \cap)^* = \{\{1,2,3\},\{2,4\},\{1,2,3,4\},\{2\}\}.$$

To obtain the closure of $\mathcal{G}$ under *relative complements,* start with the members of $\mathcal{G}$: $\{1,2,3\},\{2,4\}$ and apply the operation to them in both orders: $\{1,2,3\} - \{2,4\} = \{1,3\}$ and $\{2,4\} - \{1,2,3\} = \{4\}$. Using the newly acquired members, we obtain yet another one: $\{1,2,3\} - \{1,3\} = \{2\} = \{2,4\} - \{4\}$. Also, with the relative complements we always have $\phi = \{2\} - \{2\}$. So,

$$(\mathcal{G}, -)^* = \{\{1,2,3\},\{2,4\},\{1,3\},\{4\},\{2\},\phi\}.$$

The (simultaneous) closure of $\mathcal{G}$ under *both finite unions and relative complements* is

$$(\mathcal{G}, \cup, -)^* = \{\{1,2,3\},\{2,4\},\{1,2,3,4\},\{1,3\},\{4\},\{2\},\{1,3,4\},\phi\}.$$

The closure of $\mathcal{G}$ under *complements* is

$$(\mathcal{G}, ')^* = \{\{1,2,3\},\{2,4\},\{4\},\{1,3\}\}.$$

The (simultaneous) closure of $\mathcal{G}$ under *both finite unions and complements* is

$$(\mathcal{G}, \cup, ')^* = \{\{1,2,3\},\{2,4\},\{1,2,3,4\},\{4\},\{1,3\},\phi, \{1,3,4\},\{2\}\}.$$

Note that $(\mathcal{G}, \cup, -)^*$ and $(\mathcal{G}, \cup, ')^*$ are already closed under finite intersections. Moreover, the (simultaneous) closure of $\mathcal{G}$ under both finite unions and complements is the same as its closure under both finite intersections and complements (as well as under unions, intersections and complements).

Care should be taken to distinguish between **simultaneous closure** and **sequential closure.** For example, $(\mathcal{G}, \cup, ')^*$ is the

simultaneous closure of $\mathcal{G}$ under both finite unions and complements. Here, the order of application of the operations is
unimportant, since all new additions to the family are to be
used along with older members under *both* the union and the
complement operations. On the other hand, the closure of
$(\mathcal{G},\cup)^*$ under complementation starts with $(\mathcal{G},\cup)^* =$
$\{\{1,2,3\},\{2,4\},\{1,2,3,4\}\}$ and seeks to close it under *complements
only:* $((\mathcal{G},\cup)^*,')^* = \{\{1,2,3\},\{2,4\},\{1,2,3,4\},\{4\},\{1,3\},\phi\}$, a family
which is smaller than $(\mathcal{G},\cup,')^*$ by two sets. The union is no
longer used on the result of complementation, and without it
$\{1,3,4\}$ and therefore $\{2\}$ cannot be obtained.

The procedure illustrated in the preceding examples on a
finite universal set does not permit any unnecessary set into the
family. In fact, if any of the added sets is removed, the
remaining family is no longer closed. That is why it is correct to
refer to the closure as *smallest*. This is even more important in
the case in which the universal set is infinite. Here, the process
of adding new members to the family may never end. (Obtain-
ing Σ^* from Σ is an example, as there are always longer strings
obtainable by concatenation.) In this case, it may be necessary
to think of the closure of a family as *the smallest family, which
contains the generating family, and which is closed under the
operation(s)*.

As an example, consider the family of all singleton sets of
positive integers as the generating family: $\{\{1\},\{2\},\{3\}, \ldots \}$. The
closure of this family under binary unions (or finite unions—it
is the same) is the family of all finite nonempty subsets of
positive integers. Incidentally, this latter set may be *defined* as
the closure of the family of singletons of positive integers under
binary unions.

**4.14 CLOSURE UNDER COUNTABLE
UNIONS AND INTERSECTIONS

The countable unions and intersections permit closures not
attainable by the finite union and intersection. To illustrate, we
$\mathbb{P}$ consider three families of subsets of the set $\mathbb{P}$ of the positive
integers.

**This section requires considerable mathematical maturity on the part of the
reader. It is also an unlikely collection of topics to be frequently encountered.

EXAMPLE 1.

$\mathcal{A} = \{$All nonempty finite subsets of $\mathbb{P}\}$.

The union of any finite number of members of $\mathcal{A}$ is itself a member of $\mathcal{A}$. Thus, $\mathcal{A}$ is closed under finite unions. Note that $\bigcup_{i=1}^{\infty} \{i\} = \mathbb{P}$, which is not a finite set; yet $\{i\}$ is a member of $\mathcal{A}$, for each positive integer i. Thus, $\mathcal{A}$ is not closed under countable unions. Note also that $\mathcal{A}$ is not closed under finite intersection, since $\{1\} \cap \{2\} = \phi$.

EXAMPLE 2.

$\mathcal{B} = \{$All finite subsets of $\mathbb{P}\}$.

$\mathcal{B}$ consists of ϕ and all the members of $\mathcal{A}$ of Example 1, above. $\mathcal{B}$ is closed under both finite unions and finite intersections. However, just like $\mathcal{A}$, $\mathcal{B}$ is not closed under countable unions, by the same example. Note that $\mathcal{B}$ is closed under relative complements: If A is a finite set of positive integers, removing any positive integer from A still leaves a finite set of positive integers. (In fact, if we start with the singleton subsets of $\mathbb{P}$ as a generating family, its simultaneous closure under finite unions and relative complements is precisely the family $\mathcal{B}$.)

A family of sets which is closed under relative complements and finite unions is called a **ring of sets.**[1] Such a family is necessarily closed under finite intersections: If A and B are members of the family, then so are $A - B$ and $A - (A - B) = A \cap B$.

EXAMPLE 3.

$\mathcal{C} = \{$All subsets of $\mathbb{P}$ which are finite or have finite complements$\}$.

$\mathcal{C}$ is closed under finite unions and under complementation (and therefore under finite intersections). Note that $\mathbb{P}$ is *not* a member of $\mathcal{B}$, but it *is* a member of $\mathcal{C}$. The reason is that ϕ, which is a member of both, yields the universal set under complementation ($\phi' = \mathbb{P}$), but not under relative complements, and there is no other way to obtain $\mathbb{P}$ with just the relative complements.

A family of sets which is closed under complements and finite unions (and therefore under finite intersections) is called a **field of sets.**[2]

---

[1] A ring of sets is also defined in the literature as a family of sets closed under both finite unions and finite intersections. Such a family need not be closed under relative complements. There are further variations to be found in the literature in the definition of a ring of sets.

[2] See Footnote 2 on page 62.

The set EVENP of all even positive integers is not a member of $\mathcal{C}$, since neither it nor its complement (ODDP—the set of all odd positive integers) is finite. The *finite* union is insufficient to reach either of these two sets from inside the family $\mathcal{C}$. However, the *countable* union can do so, since EVENP $= \bigcup_{i=1}^{\infty} \{2i\}$.

The countable union is obviously a "stronger" operation than the finite union. A family of sets closed under countable unions and relative complements (and therefore under countable intersections) is called a σ-**ring of sets**[1] (pronounced, "sigma-ring"). Since closure under countable unions implies closure under finite unions, any σ-ring of sets is also a ring of sets. A family of sets closed under countable unions and complements (and therefore also under countable intersections) is called a σ-**field of sets.**[2] Again, a σ-field of sets is also a field of sets, for a similar reason.

The following is a summary of the definitions and a terse collection of elementary properties of the families of sets mentioned.

4.22 Definition: Let $\mathcal{F}$ be a nonempty family of subsets of a universal set U. Consider the following closure properties of $\mathcal{F}$:

(i) $\forall A_1, A_2 \in \mathcal{F}, A_1 \cup A_2 \in \mathcal{F}$.   (*Binary union*)

(ii) $\forall A_1, A_2 \in \mathcal{F}, A_1 - A_2 \in \mathcal{F}$.   (*Relative complement*)

(iii) $\forall A \in \mathcal{F}, A' \in \mathcal{F}$.   (*Complement*)

(iv) If $A_i \in \mathcal{F}$ for each positive integer i, then $\bigcup_{i=1}^{\infty} A_i \in \mathcal{F}$.   (*Countable union*)

The family $\mathcal{F}$ is a **ring**[3] (**Boolean ring**) **of sets** iff it has properties (i) and (ii). The family $\mathcal{F}$ is a **field** (**algebra, Boolean algebra, additive class**) **of sets** iff it has properties (i) and (iii). The family $\mathcal{F}$ is a σ-**ring**[3] **of sets** iff it has properties (ii) and (iv). The family $\mathcal{F}$ is a σ-**field** (σ-**algebra, completely additive class**) **of sets** iff it has properties (iii) and (iv). ∎

---

[1]A σ-ring of sets is also defined in the literature as a family of sets closed under countable unions and countable intersections. Such a family need not be closed under relative complements. There are further variations to be found in the literature in the definition of a σ-ring of sets.

[2]There are variations to be found in the literature in the definitions of a field of sets and a σ-field of sets. The reader is cautioned to verify the definition used by the specific author in each case.

[3]The reader should be warned that there is no unanimity in the literature with

EXAMPLE 4:
For any universal set U, $\{\phi\}$ is the smallest ring (and σ-ring). The power set $\mathcal{P}(U)$ is the largest ring, field, σ-ring and σ-field. $\{\phi, U\}$ is the smallest field (and σ-field).

EXAMPLE 5:
Let U be any infinite set. The family $\mathcal{C}$ of all subsets of U which are finite or have finite complements (see example 3, above, for a special case) is a field, but is not a σ-field. The family $\mathcal{D}$ of all infinite subsets of U is none of the above.

EXAMPLE 6:
ℝ

Let $U = \mathbb{R}$, the set of real numbers, and let $\mathcal{E}$ be the family of four subsets ϕ, $\{x: x < 0\}$, $\{x: x \geq 0\}$ and U. Then $\mathcal{E}$ is a σ-field.

EXAMPLE 7:
Let U be an infinite set, and let $\mathcal{F}$ be the family of all sets which are countable or have countable complements. Then $\mathcal{F}$ is a σ-field.

4.23 Theorem:

a. Any ring of sets is closed under finite intersection.

b. Any ring of sets includes ϕ.

c. Let $\mathcal{F}$ be a ring of sets, and let $A_i \in \mathcal{F}$ for each $i \in \{1,2,\ldots,n\}$, for a positive integer n. Then both $\bigcup_{i=1}^{n} A_i$ and $\bigcap_{i=1}^{n} A_i$ are in $\mathcal{F}$.

d. A ring of subsets of U is a field if and only if it contains U.

e. Any field of sets has property (ii); i.e., if $\mathcal{F}$ is a field and $A,B \in \mathcal{F}$, then $(A' \cup B)' = A - B \in \mathcal{F}$.

f. A ring is a σ-ring iff it is closed under countable unions.

g. A σ-ring is a σ-field iff it contains the universal set U.

h. A σ-ring is closed under countable intersections.

i. Every σ-ring is a ring.

j. Every σ-field is a field.

k. Every σ-field is a σ-ring.

---

regard to the definitions of these terms. A ring of sets is also defined as a family closed under finite unions and finite intersections, and a σ-ring as a family closed under countable unions and countable intersections. Other definitions also exist for these two terms, as well as for a field of sets and for a σ-field of sets. Care should be taken in each case to verify the intention of the particular author.

4.15 SET PRODUCT; MORE ABOUT STRINGS

[This section is a continuation of Section 4.10. The concepts and definitions of Section 4.10 are assumed here. They include **alphabet, strings over an alphabet, concatenation of strings, length of a string, closure** of a set, language, the **empty language** ϕ and the language $\Lambda = \{\lambda\}$.]

In many programming languages, a variable name must start with a letter and may then be followed by a number or characters, each of which is a digit or a letter.

Denote here the set of letters by L, the set of digits by D, and the union of the two by C (characters which are letters or digits). Thus,

$$L = \{A,B, \ldots ,Z\}, D = \{0,1, \ldots ,9\},$$
$$C = \{A,B, \ldots ,Z,0,1, \ldots ,9\}.$$

A variable name of two characters must then be the *concatenation* of a member of L and a member of C. The set of all variable names of two characters consists of all such concatenations:

AA, AB, . . . , AZ, A0, A1, . . . , A9

BA, BB, . . . , BZ, B0, B1, . . . , B9

. . . .

ZA, ZB, . . . , ZZ, Z0, Z1, . . . , Z9 .

This set of strings may be described as $\{xy: x \in L$ and $y \in C\}$.

It should be clear that the list consists of **strings of length** 2, the first character of which comes from L and the second from C, and *the order must not be reversed*. We call such a set the **product**, or **set product**, of L and C and denote it by $L \cdot C$, or just by LC.

> *4.24 Definition:*   Let L and C be sets. The **set product** (alternately, **product**) of L and C is denoted by $L \cdot C$, or LC and is defined by

$L \cdot C$ $$L \cdot C = \{xy: x \in L \text{ and } y \in C\}.$$

The set product is a binary operation on sets. It is a very useful tool in a variety of concerns in computer science. It differs from the Cartesian product of two sets (see Section 7.4) mainly in that the elements of the set product are *strings*, rather than ordered pairs of elements.

To continue our example, we can obtain all variable names of length 3 by forming the product of the set $L \cdot C$ (of variable names of length 2) with the set C; that is, $(L \cdot C) \cdot C$. The resulting set includes such strings as B6K and Z17.

It should be easy to see that the same result is obtained from the product of L with $C \cdot C$—a letter followed by two characters, each of which is either a letter or a digit. This, in fact is a property of the set product: the grouping is unimportant, as long as the order remains the same.

> **4.25 Theorem:** The set product is associative. That is, where A, B and C are sets,
>
> $$(A \cdot B) \cdot C = A \cdot (B \cdot C).$$

As a consequence of this theorem, it makes sense to drop the parentheses altogether and write $A \cdot B \cdot C$ or ABC.

When the product is of a set with itself, especially when the number of factors is large, it is convenient to use the notation of exponents. Thus, $C \cdot C$ may be written as C^2, CCC as C^3, etc. It is then convenient to denote the "product of one factor of C," that is just C itself, as its own first power: $C^1 = C$. (The purist may note that C^1 is the set of all strings of length 1 of members of C, while C is the set whose elements are used to create the *strings* which are the members of C^1. This distinction is ignored in the rest of this treatment.)

The 0-th power of C, or of *any* set, is best defined as the set $\Lambda = \{\lambda\}$ whose only member is the empty string.

> **4.26 Definition:** Let A be a set. We define the **powers** of A as,
>
> $$A^0 = \Lambda = \{\lambda\},$$
> $$A^1 = A,$$
> $$A^2 = A \cdot A = \{xy: x \in A \text{ and } y \in A\}.$$

A^0

For any nonnegative integer n,

$$A^{n+1} = A^n \cdot A.$$

A^n

In our example, if the length of a variable name were restricted to between 1 and 4, the set of all allowable variable names could be described as

$$L \cup LC \cup LC^2 \cup LC^3.$$

This same set could be described more economically if we used the set $C^\lambda = C \cup \lambda$, whose elements are all the letters, all the digits and the *empty string* λ. The product $L \cdot C^\lambda$ still includes all the concatenations of elements of L and those of C (the members of $L \cdot C$). But, in addition, $L \cdot C^\lambda$ includes all concatenations of members of L with λ: $A\lambda = A$, $B\lambda = B$, etc. The result is then

$$L \cdot C^\lambda = L \cdot (C \cup \{\lambda\}) =$$
$$L \cdot (C \cup \lambda) = L \cdot C \cup L \cdot \lambda = L \cdot C \cup L;$$

that is, all variable names of length 1 and 2.

In a similar fashion, we could express the set of all variable names of length ≤ 4 as

$$L(C^\lambda)^3$$
$$= L(C \cup \lambda)(C \cup \lambda)(C \cup \lambda)$$
$$= (LC \cup L)(C \cup \lambda)(C \cup \lambda)$$
$$= (LCC \cup LC\lambda \cup LC \cup L\lambda)(C \cup \lambda)$$
$$= LCCC \cup LCC\lambda \cup LCC \cup LC\lambda \cup LC \cup L\lambda$$
$$= LC^3 \cup LC^2 \cup LC \cup L.$$

The property $L(C \cup \lambda) = LC \cup L\lambda$ is true in the general case: *the set product distributes over the union.*

4.27 Theorem: Let A, B and C be sets. Then,
$$A(B \cup C) = AB \cup AC.$$

The reader should note that the notation of exponents should be used with the same care as it is used for the customary exponents: LC^2 is different from $(LC)^2$. The former produces strings of length 3, while the strings of the latter are of length 4. For example, $A2T6 \in (LC)^2$; also, A and $A25$ are elements of $L(C^\lambda)^2$ but not of $(LC^\lambda)^2$.

Two basic results are important to remember and to contrast with each other. The first concerns the product with the empty set:

$$A \cdot \phi = \{xy: x \in A \text{ and } y \in \phi\}.$$

By definition of the product (Definition 4.24), the strings in $A \cdot \phi$ must have an element from A followed by an element from ϕ. However, ϕ has no elements and hence no such string xy can be formed. The conclusion is then that $A \cdot \phi$ has no elements and is therefore the empty set; that is, $A \cdot \phi = \phi$.

On the other hand, $A \cdot \Lambda = A \cdot \{\lambda\} = \{xy: x \in A \text{ and } y = \lambda\} = \{x\lambda: x \in A\} = \{x: x \in A\} = A$. We thus have,

4.28 Theorem: Let A be a set. Then,

(i) $A\phi = \phi A = \phi$;

(ii) $A\Lambda = \Lambda A = A$.

It should be noted, in contrast to Theorem 4.27, that *the set product does not distribute over intersection:* Let $A = \{a,ab\}$ and $B = \{b\}$. Then $A(B \cap \Lambda) = A\phi = \phi$, while $AB = \{ab,abb\}$, $A\Lambda = A = \{a,ab\}$ and hence $AB \cap A\Lambda = \{ab\}$.

It should also be noted that *the set product is not commutative:* $\{a\}\{b\} = \{ab\}$ while $\{b\}\{a\} = \{ba\}$.

The following is a collection of results concerning the set product and its interaction with other basic set operations. (Some of these results have already appeared above.)

4.29 Theorem: Let A, B and C be sets and let m and n be arbitrary nonnegative integers.

(a) If $A \subseteq B$ and $C \subseteq D$, then $AC \subseteq BD$.

(b) $A(B \cup C) = AB \cup AC$; $(B \cup C)A = BA \cup CA$.

(c) $A(B \cap C) \subseteq AB \cap AC$; $(B \cap C)A \subseteq BA \cap CA$.

(d) $A^m A^n = A^{m+n}$.

(e) $(A^m)^n = A^{mn}$.

(f) If $A \subseteq B$ then $A^m \subseteq B^m$.

4.16 SET CLOSURE REVISITED; YET MORE ABOUT STRINGS

[This section uses the concepts, definitions and some of the results of Sections 4.10, 4.11 (countable unions), 4.13 and 4.15. The concept of **closure** (Kleene closure) is central to this section.]

The set A^* (read "A star") of all strings over the alphabet A was defined in Section 4.10 (Definition 4.9), and was named the **closure** (or the **Kleene closure**) of A. (It is sometimes called the **star closure** of A.) It was also depicted in Section 4.13 as the smallest set (of strings) which includes λ and the members of A and which is closed under arbitrary finite concatenation.

Here we present another view of the closure of a set which, in a way, is more constructive. Since A^* consists of all strings of finite length of members of A, such strings may be arranged in order of their length; the strings of each length occupy a separate set. The first of these sets has as its only member λ, the one and only string of length 0. This set may be written as A^0 or as Λ.

Next comes the set of all strings of length 1, which is A itself. It may be written as A^1. The next sets in order are the set A^2 of all strings of length 2, the set A^3 of all strings of length 3, and so on *without end*. There is no end to the procession of such powers of A because concatenation produces arbitrarily long strings over A. Our claim is that A^* is precisely the countable union (see Section 4.11) of all these sets:

4.30 Theorem: Let A be an alphabet. Then,

A^*
$$A^* = \bigcup_{i=0}^{\infty} A^i = A^0 \cup A^1 \cup A^2 \cup \ldots \cup A^n \cup \ldots$$

(In deciphering the theorem, it may help to recall that $A^0 = \Lambda = \{\lambda\}$, $A^1 = A$, and $A^{n+1} = A^n \cdot A$, by Definition 4.26 of Section 4.15.)

The reasoning behind Theorem 4.30 is of some importance in understanding its applications, and so we sketch it here: Any member of A^* is a string of finite length, say k, over A. It must therefore be in A^k and hence in $\bigcup_{i=0}^{\infty} A^i$, specifically in its $k + 1$st component. On the other hand, any member of the countable union must be a member of one of its components and hence a string of specific length over A. It is therefore a member of A^*. The equality follows by double inclusion (Law 2 of Table 4.1, Section 4.3).

A set similar to A^*, also frequently used in computer science, is the **positive closure** of A, denoted A^+ (read "A plus"). It consists of all strings of *positive length* of members of A.

4.31 Definition: Let A be an alphabet. Then, the **positive closure** of A is

A^+
$$A^+ = \bigcup_{i=1}^{\infty} A^i = A^1 \cup A^2 \cup \ldots \cup A^n \cup \ldots$$

The following results are immediate from the definitions.

4.32 Theorem: For any set A,

(i) $A^* = A^+ \cup \Lambda$.

(ii) $A^n \subseteq A^+$, for every positive integer n.

(iii) $A^n \subseteq A^*$, for every nonnegative integer n.

An important fact, easy to overlook, is that *the empty string λ is a member of every closure:*

4.33 Theorem: For any set A, $\lambda \in A^*$.

This applies to the empty set as well: $\phi^* = \phi^0 \cup \phi^1 \cup \phi^2 \cup \ldots = \Lambda \cup \phi \cup \phi \cup \ldots = \Lambda = \{\lambda\}$.

On the other hand, the positive closure of the empty set is itself empty:

$$\phi^+ = \phi.$$

The following is a collection of results concerning the closure and the positive closure of sets.

4.34 Theorem: Let A and B be sets. Then,

(a) $(A^+)^+ = A^+$.

4.34 Theorem: —*Continued*

(b) $(A^*)^* = A^* = A^*A^*$.

(c) If $A \subseteq B$, then $A^+ \subseteq B^+$ and $A^* \subseteq B^*$.

(d) $A^+ = A^1 \cup A^2 \cup \ldots \cup A^{n-1} \cup A^n \cup A^nA^+$, for any positive integer n.

(e) $A^* = \Lambda \cup A^1 \cup A^2 \cup \ldots \cup A^{n-1} \cup A^nA^*$, for any nonnegative integer n.

(f) $A^+ = AA^* = A^*A$.

(g) $\Lambda^+ = \Lambda = \Lambda^*$.

(h) $AA^+ = A^+A$.

(i) $A \subseteq AB^*$ and $A \subseteq B^*A$.

(j) $(A^*)^+ = (A^+)^* = A^*$.

(k) $A^*A^+ = A^+A^* = A^+$.

The fact should not be overlooked that a set of strings itself may serve as an "alphabet" and may be involved in the product and closure operations. When such is the case, it is possible for the positive closure of A to contain λ as an element. For example, if $A = \{\lambda, a\}$ then $A^+ = A^*$. In fact,

4.35 Theorem: Let A be a set of strings. Then,

$$\lambda \in A \text{ iff } A^+ = A^*.$$

One of the more useful results in manipulating sets of strings (and, consequently, in regular expression manipulation) is an easy consequence of the previous ones. It is sufficiently useful to be given prominence as a theorem.

4.36 Theorem: Let A be a set. Then,

$$AA^* \cup \Lambda = A^*.$$

The following is a further collection of results concerning set closure; most of them are easy consequences of the definitions and the preceding results.

4.37 Theorem: Let A and B be sets. Then,

(a) $A(BA)^* = (AB)^*A$.

(b) $(A^*B^*)^* = (A \cup B)^* = (A^* \cup B^*)^* = (A^* \cup B)^*$.

4.37 Theorem —*Continued*

(c) $(A \cup B)^* = A^*(BA^*)^*$.

(d) $(A^*B^*)^* = (B^*A^*)^*$.

(e) $(A \cup B^*)^* = (B \cup A^*)^*$.

The reader should be careful to avoid the following traps. The counterexample $A = \{a\}$, $B = \{b\}$, $C = \{c\}$ may be used to illustrate for each that it is false.

$AB^* = A^*B$ is false: $abb \in AB^* - A^*B$.

$AB = BA$ is false: $ab \in AB - BA$.

$A^*B^* = B^*A^*$ is false: $ab \in A^*B^* - B^*A^*$.

$(AB)^* = (BA)^*$ is false: $ab \in (AB)^* - (BA)^*$.

$A^*(BA) = (AB)A^*$ is false: $aaba \in A^*(BA) - (AB)A^*$.

$A(B \cup C)^* = AB^* \cup AC^*$ is false: $abc \in A(B \cup C)^* - (AB^* \cup AC^*)$.

$A(B \cup C)^* = (AB \cup AC)^*$ is false: $abac \in (AB \cup AC)^* - A(B \cup C)^*$.

We conclude the section with a celebrated result, often used in finite automata and formal language theory. It provides the solution of a set-equation in the set-variable X in terms of the constant sets A and B.

4.38 Theorem (Dean Arden): Let A and B be sets of strings (over an alphabet) and let $\lambda \notin A$. Then, the equation $X = AX \cup B$ has the unique solution $X = A^*B$, and the equation $X = XA \cup B$ has the unique solution $X = BA^*$.

As an example, let $\Sigma = \{0,1\}$, let A be the set of all binary strings of even positive length over Σ and let $B = \Sigma = \{0,1\}$. Then, the equation $X = AX \cup B$ has the unique solution $X = A^*B$, which is the set of all binary strings of odd length. (Note that $A^* = \Lambda \cup A$ in this case.)

5

Discrete Probability and Combinatorial Analysis

[The reader may wish to review Section 4.9 before reading this chapter. The study of discrete probability is a frequent and useful tool to the computer scientist, and this is the main reason for including Chapter 5 in this book. However, this chapter is presented also as an application of the algebra of sets and, in particular, of the principle of inclusion and exclusion of Section 4.9.]

5.1 BACKGROUND: THEORY VS. PRACTICE

The historical purpose of the study of probabilities was to imitate reality closely enough so that the theoretical results could guide the making of practical decisions. (Originally, these were decisions in gambling.) This is still true in the way we *use* probabilities: We seek the *expectations,* the *likelihoods;* we want to know "the odds."

In practice, we may "go with the odds" or defy them and "play a hunch." If we defy the odds, it is perhaps because they promise accuracy only over a large number of trials, but are not expected to be particularly helpful in a single trial. The "one in a million" shot is expected to happen once in a million trials, but it could happen the very next time—just ask any addicted gambler. Moreover, when it does happen, it no longer matters how likely it was to happen.

Whether probabilities form a good basis for decision making may depend on a variety of factors. The application of the theory still strives to imitate reality as closely as possible (whatever that means), and for that purpose, certain assumptions must be made. One basic assumption, in the absence of evidence to the contrary, is that *each of the basic possibilities has an equal chance of happening.*

For example, the common die with six faces has six possible outcomes when it is thrown (or rolled):

We assume each to have "1 out of 6" (that is, 1/6) chance of showing at the top of the die when it comes to rest. We then assign the "odds" or **probability** of 1/6 to each face. However, if the substance of the single dot on the 1-face is extremely massive, it is more likely to pull the 1-face down to the bottom and the opposite 6-face is more likely to show at the top. Such "loaded dice" change the actual odds; to imitate them more closely, different probabilities must be assigned to the various faces.

The rules for assigning such probabilities are given below. They are rules for the *theory* only, and are intended only to help decision making, not to guarantee success. They are to be applied when more than one outcome is possible in an "experiment." Such a situation is often referred to as **random,** although the term is more often used to describe equal likelihood of happening. The reader may safely assume **random selection** of alternatives to mean that each alternative has an equal chance of being selected.

5.2 SAMPLE POINTS, SAMPLE SPACE, EVENTS AND PROBABILITY

To continue with the example of the die of the preceding section, each of the six faces of the die is called a **sample point.** It is the basic single outcome of the experiment. The collection of all six faces, of all the sample points, is called the **sample space,** often denoted by S. In a real sense, the sample space is the "universal set" of concern here. We wish to fix the odds for a certain subset of S, to determine "how likely" is a particular collection of faces or a succession of them.

For example, with a perfectly balanced die, we wish for all faces to have equal probabilities, so the odds that one throw of the die results in a face that is less than 3 should be fixed as $1/3$. The subset "less than 3" of S is called an **event.** It includes the sample points 1 and 2 and, therefore, has two chances out of six to occur. That is why the probability assigned to the event is $2/6 = 1/3$.

Each subset of S is called an **event.** The singleton subsets of S, those including one sample point each, are called **simple events,** and they are the events to which the original probabilities are assigned.

$p(E)$ We denote the **probability of an event** E by $p(E)$. Thus, abbreviating "the face showing after the throw is 3" by just "3," we say that $p(3) = 1/6$. Similarly, the probability that the face showing is less than 3 is $p(< 3) = 1/3$. If E denotes the event of an even face, then $p(E) = p(2 \text{ or } 4 \text{ or } 6) = 1/2$. Likewise, $p(\text{not } 4) = 5/6$ and $p(\text{odd and less than } 5) = p(1 \text{ or } 3) = 1/3$.

The term **discrete probability** is a description of the size of the sample space: it is either finite or "countably infinite" (like the set of positive integers of Sections 15.3 and 15.4). In this coverage, we restrict our attention to the discrete case, since the continuous one is not likely to confront the computer scientist and the applied mathematician.

5.3 COMBINATIONS OF EVENTS

We return to our example and let E be the event "an even face" and F the event "less than 5." Then $E = \{2,4,6\}$ and $F = \{1,2,3,4\}$. The event "an even face or less than 5" may be written as "E or F" and computed as $E \cup F = \{1,2,3,4,6\}$. Thus, $p(E \cup F) = 5/6$. Now, $p(E) = 1/3$ and $p(F) = 2/3$. Their sum is not $5/6$. For the sake of clarity, we use a modified Venn diagram, with the two circles denoting the regions representing the events E and F, respectively. We then place the various faces of the die in the correct regions of the diagram, as illustrated in Figure 5.1. Each face of the die is placed in a unique subregion of the diagram, determined by E

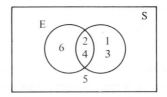

The regions determined by E and F.

FIGURE 5.1

and F. The event $E \cup F$ is then split into the *mutually exclusive* regions $E - F = \{6\}$, $F - E = \{1,3\}$ and $E \cap F = \{2,4\}$.

Now we can *add probabilities* for the mutually exclusive events:

$$p(E \cup F) = p(E - F) + p(F - E) + p(E \cap F)$$
$$= 1/6 + 2/6 + 2/6 = 5/6.$$

The situation here is the same as in the principle of inclusion and exclusion of Section 4.9:

$$\#(E \cup F) = \#(E) + \#(F) - \#(E \cap F) \text{ and so}$$
$$p(E \text{ or } F) = p(E) + p(F) - p(E \text{ and } F) =$$
$$3/6 + 4/6 - 2/6 = 5/6.$$

In fact, one of the rules (axioms) for discrete probabilities may be regarded as an application of the rule basic to the principle of inclusion and exclusion, which we paraphrase for probabilities as follows:

The probability of the OR-combination of mutually exclusive events is the sum of their respective probabilities.

In symbols, if A and B are events such that $A \cap B = \phi$, then $p(A \cup B) = p(A) + p(B)$. As a result, all of Theorem 4.6 is immediately applicable, with the obvious translation of OR for $\cup$, AND for $\cap$, NOT for $'$, A but not B for $A - B$, and p for $\#$. This translation appears as Theorem 5.5, below.

5.4 PROBABILITY DEFINED

There are two additional rules for probabilities. The first aims to standardize the computations, so that all probabilities are fractions of 1; it fixes the largest probability, that of the entire sample space, at 1,

$$p(S) = 1.$$

This may be interpreted to mean that the experiment must result in some sample point every time the experiment is conducted. This also forces the *sum* of the probabilities of all simple events (sample points) to be 1. It has the effect of expressing "*k* out of *n* chances" as the fraction k/n. That is, the probability assigned to an event E is *the number of favorable cases divided by the number of all possible cases.* In other words, the proportion of the number of sample points in E to the total number of sample points in the sample space S.

The second rule prohibits negative probabilities, for obvious intuitive reasons. Thus, for any event E, $p(E) \geq 0$. This has the effect

$$0 \leq p(E) \leq 1.$$

> **5.1 Definition:**   Let S be a (countable) set, called the **sample space.** To each subset E of S, called an **event,** assign a real number $p(E)$. (This makes p a real-valued function on $\mathcal{P}(S)$.) Then these numbers are **probabilities** if and only if
>
> (i) $p(E) \geq 0, \forall E \subseteq S$;
> (ii) $p(S) = 1$;
> (iii) $\forall A,B \subseteq S$, if $A \cap B = \phi$ then
> $\qquad p(A \cup B) = p(A) + p(B)$.

5.5 BASIC RESULTS

We assume S to be a discrete sample space in all cases. The following results are immediate from the definition and the algebra of sets.

> **5.2 Theorem:**   Let $E_1, E_2, \ldots$ be the simple events of S (the singleton subsets). Then,

$$\sum_{i=1}^{\infty}$$

$$p(E_1) + p(E_2) + \ldots = \sum_{i=1}^{\infty} p(E_i) = 1.$$

If S is finite and the simple events are $E_1, E_2, \ldots, E_n$ then

$$\sum_{i=1}^{n}$$

$$\sum_{i=1}^{n} p(E_i) = 1.$$

5.3 Theorem:　Let A be any event in S. Then $0 \le p(A) \le 1$.

5.4 Theorem:　Let $A_1, A_2, \ldots, A_n$ be mutually exclusive events. Then

$$p(\bigcup_{i=1}^{n} A_i) = \sum_{i=1}^{n} p(A_i) = p(A_1) + p(A_2) + \ldots + p(A_n).$$

The last of the sequence of results immediate from previous definitions and results to be presented here is the translation of Theorem 4.8.

5.5 Theorem:　Let S be a sample space, let A and B be any two events in S and let ϕ be the empty event (i.e., the event with no sample points).

1. a.　$p(\phi) = 0$;
   b.　$p(A \cup B) = p(A) + p(B) - p(A \cap B)$;
   c.　$p(A - B) = p(A) - p(A \cap B)$;
   d.　$p(B') = 1 - p(B)$.
2. If A and B are mutually exclusive, i.e. $A \cap B = \phi$, then

   a.　$p(A \cap B) = 0$;
   b.　$p(A \cup B) = p(A) + p(B)$;
   c.　$p(A - B) = p(A)$.
3. If B implies A, i.e. A must occur if B occurs, then

   a.　$p(B) \le p(A)$;
   b.　$p(A \cup B) = p(A)$;
   c.　$p(A \cap B) - p(B)$;
   d.　$p(A - B) = p(A) - p(B)$;
   e.　$p(B - A) = 0$.

5.6 EXAMPLE: THROWING TWO DICE

When two (familiar) dice are thrown, the outcome is one of 36 possible double faces: one die shows one of six faces and so does the other. The sample space is then

$$S = \{(i, j): 1 \le i, j \le 6\} = \{(1,1),(1,2),\ldots,(6,5),(6,6)\}.$$

Each such pair represents a *simple event* of S. They are all assumed to be equally likely and hence each is assigned the probability $1/36$.

The following are examples of events in this sample space and their computed probabilities.

EXAMPLE 1:

Event A: The sum of the two faces is 5.

Event A includes exactly the sample points (1,4), (2,3), (3,2), (4,1) and hence $p(A) = 4/36 = 1/9$.

It is easy to compute in the same fashion the probabilities of the eleven possible totals of the faces of two dice, by counting the different ways each total can be formed and dividing this number by 36:

$$p(2) = 1/36 \qquad\qquad p(12) = 1/36$$
$$p(3) = 2/36 = 1/18 \qquad\qquad p(11) = 2/36 = 1/18$$
$$p(4) = 3/36 = 1/12 \qquad\qquad p(10) = 3/36 = 1/12$$
$$p(5) = 4/36 = 1/9 \qquad\qquad p(9) = 4/36 = 1/9$$
$$p(6) = 5/36 \qquad\qquad p(8) = 5/36$$
$$p(7) = 6/36 = 1/6.$$

The most likely total of the faces of two dice is 7, but it is more likely than 6 or 8 by just a little (1/36).

EXAMPLE 2:

Event B: Both faces are odd.

There are nine such pairs: (1,1), (1,3), ..., (5,3), (5,5). Hence, $p(B) = 9/36 = 1/4$. (We return to this case in the discussion of independent events in Section 5.8.)

EXAMPLE 3:

Event C: The face of the first die is smaller than that of the second. (Here, we must recognize one die as the first. We could consider one red die and one blue, and the first is the red die.)

We use Theorem 5.4 and decompose event C into the six mutually exclusive subevents C_i = The second die shows the face i, for $i \in \{1, \ldots, 6\}$. (The face of the first die is still smaller than the second.) We then count the number of points in each subevent and compute the probabilities accordingly:

$C_1 = \phi$, since the face of the first die cannot be less than 1. So, $p(C_1) = 0$.

$C_2 = \{(1,2)\}$ and so $p(C_2) = 1/36$.

$C_3 = \{(1,3),(2,3)\}$ and so $p(C_3) = 2/36 = 1/18$.

Similarly, $p(C_4) = 3/36$, $p(C_5) = 4/36$ and $p(C_6) = 5/36$.

Hence,

$$p(C) = \sum_{i=1}^{6} p(C_i)$$

$$= (0 + 1 + 2 + 3 + 4 + 5)/36 = 15/36 = 5/12.$$

EXAMPLE 4:

Event D: The total is even.

Again, by decomposing event D into mutually exclusive subevents and adding their probabilities (listed in Example 1, above), we get

$$p(D) = p(2) + p(4) + p(6) + p(8) + p(10) + p(12)$$

$$= (1 + 3 + 5 + 5 + 3 + 1)/36 = 18/36 = 1/2.$$

Note that it is no longer necessary to compute in detail the probability that the total is odd, because this is the event D', the complement of D: By Theorem 5.5 1d,

$$p(D') = 1 - p(D) = 1 - 1/2 = 1/2.$$

EXAMPLE 5:

Event E: The total is at least 4.

It would be somewhat cumbersome to compute $p(E) = p(4) + p(5) + \ldots + p(12)$. Instead, we compute $p(E) = 1 - p(E')$, since E' is the event "the total is less than 4." Then $E' = \{(1,1),(1,2),(2,1)\}$ and $p(E') = 3/36 = 1/12$. Hence, $p(E) = 1 - 1/12 = 11/12$.

EXAMPLE 6:

Event F: The total is either even or at least 4 (or both). $F = (D$ or $E)$; that is, $F = D \cup E$. Consequently, $p(F) = p(D) + p(E) - p(D \cap E)$. The event $D \cap E = (D$ and $E)$ may be decomposed into the totals 4, 6, 8, 10 and 12, and hence $p(D \cap E) = (3 + 5 + 5 + 3 + 1)/36 = 17/36$. Consequently,

$$p(F) = 1/2 + 11/12 - 17/36 = 17/18.$$

The same result may be reached by decomposing F into the mutually exclusive events E and "the total is 2," and then

$$p(F) = 11/12 + 1/36 = 17/18.$$

5.7 CONDITIONAL PROBABILITY

In many cases, the interesting question is not "How likely is event A?" but "How likely is event A, once we know that event B happened?"

Suppose, for example, that the frequency of the letters in a page of text is used to fix for each letter the probability that it appears in a word chosen at random from a book. The probability that "q appears in the word" is likely to be relatively small, since q is not a frequently used letter. But if we already know

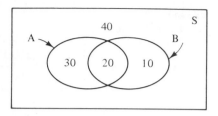

Illustrating conditional probability

FIGURE 5.2

that u appears in the word, the probability that q also appears in the word should be higher. That is, if we looked only at words with u in them, we would meet q more often than we would ordinarily on a page.

The concept of "the probability that A occurs, given that B occurred" may be illustrated by the diagram of Figure 5.2. In it, 100 objects, or sample points, are distributed so that 40 are outside both events A and B, 20 are in both events, 30 are in A but not in B, and 10 are in B but not in A. Knowing, or assuming, that B has occurred is tantamount to regarding B as the universal set and ignoring its complement altogether. Then, the part of B that is also in A has 20 objects and they are the favorable cases. All possible cases are the 30 objects of B, and hence the probability we seek is the ratio $20/30 = 2/3$. By contrast, the (absolute) probability of A is the ratio of the number 50 of objects in A to the total number 100 of objects in S; that is, $50/100 = 1/2$.

The diagram correctly illustrates the concept and the basis for both the definition and the calculations of **conditional probabilities:** the probability $p(A|B)$ of A on the condition that B occurred is the probability $p(A \cap B)$ that both occurred divided by the probability $p(B)$ that B occurred; that is,

$$\frac{20/100}{30/100} = \frac{2}{3}.$$

5.6 Definition: Let A and B be two events in the sample space S, such that $p(B) > 0$. Then, the (**conditional**) **probability of A on the condition B** (also, **the probability of A, given that B has occurred**) is denoted by $p(A|B)$ and is defined as

$$p(A|B) = \frac{p(A \cap B)}{p(B)}.$$

Note that, by Definition 5.6, $p(B|B) = 1$, as it should be. Also note that $p(A \cap B) = p(A)p(A|B)$. In fact, this can be generalized to the following.

5.7 Theorem: Let A_i be an event in the sample space S, for each $i \in \{1, \ldots, n\}$. Then, the conditional probability of A_n, given that $A_1, \ldots, A_{n-1}$ occurred, is

$$p(A_n|A_1,A_2,\ldots,A_{n-1}) = \frac{p(A_1 \cap A_2 \cap \ldots \cap A_{n-1} \cap A_n)}{p(A_1 \cap A_2 \cap \ldots \cap A_{n-1})}.$$

Moreover,

$$p(A_1 \cap A_2 \cap \ldots \cap A_n) =$$

$$p(A_1)p(A_2|A_1)p(A_3|A_1 \cap A_2)\ldots p(A_n|A_1 \cap A_2 \cap \ldots \cap A_{n-1}).$$

When the event B of Figure 5.2, above, is partitioned into subevents B_1, B_2, B_3, B_4 (so that every pair B_i and B_j is disjoint if $i \neq j$ and the union $B_1 \cup B_2 \cup B_3 \cup B_4 = B$ exhausts B), the diagram in Figure 5.3 illustrates the anticipated relationship:

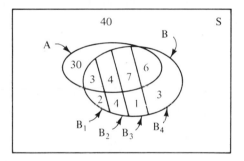

Illustrating Theorem 5.8

FIGURE 5.3

The favorable cases for "A occurs, given that B has occurred" are 3 out of the 5 of B_1, 4 out of the 8 of B_2, 7 out of the 8 of B_3 and 6 out of the 9 of B_4. They still add up to 20 out of the 30 objects of B. However, the probability may also be computed as follows:

$$p(A|B) = \frac{p(A \cap B)}{p(B)}.$$

Now, $A \cap B = A \cap \left(\bigcup_{i=1}^{4} B_i \right) = \bigcup_{i=1}^{4} (A \cap B_i)$

and, since the B_i are mutually disjoint,

$$p(A \cap B) = p\left(\bigcup_{i-1}^{4} (A \cap B_i)\right) = \sum_{i-1}^{4} p(A \cap B_i).$$

Now, for each $i \in \{1, \ldots, 4\}$, $p(A \cap B_i) = p(A|B_i)p(B_i)$ and so

$p(A|B) =$

$$\frac{p(A|B_1)p(B_1) + p(A|B_2)p(B_2) + p(A|B_3)p(B_3) + p(A|B_4)p(B_4)}{p(B)} =$$

$$\frac{(3/5)(5/100) + (4/8)(8/100) + (7/8)(8/100) + (6/9)(9/100)}{30/100} =$$

$$\frac{(3 + 4 + 7 + 6)/100}{30/100} = \frac{20/100}{30/100} = \frac{20}{30} = \frac{2}{3}$$

5.8 Theorem: Let A and B be two events in the sample space S, such that $p(B) > 0$, and let $B_1, B_2, \ldots$ be subevents of B such that $B = \bigcup_i B_i$ and $B_i \cap B_j = \phi$ if $i \neq j$. Then

$$p(A|B) = \left(\sum_i p(A|B_i)p(B_i)\right)/p(B).$$

We illustrate the use of Theorem 5.8 with the aid of the following.

EXAMPLE:
The 100 students of a rural school come to school in three buses, 25 in the first, 35 in the second and 40 in the third. Over the year, the frequency of students who forget their lunch tickets at home has been fairly stable for each pick-up region and they are now taken as probabilities: 4%, 20% and 10%, respectively. What is the probability that a student selected at random has forgotten his/her lunch ticket?

Let A denote forgetting a lunch ticket and B_1, B_2 and B_3 belonging to the corresponding pick-up region. To use Theorem 5.8, we note that $B = B_1 \cup B_2 \cup B_3$ must occur (the student must be from one of the three regions) and so $p(B) = 1$. Then.

$$p(A) = p(A|B_1)p(B_1) + p(A|B_2)p(B_2) + p(A|B_3)p(B_3)$$

$$= (.04)(.25) + (.20)(.35) + (.10)(.40) = 0.12.$$

We conclude this section with a result which is helpful in many applications. It is sometimes called **Bayes' rule for the probability of causes,** since the B_i's of the preceding example may be regarded as causes in similar cases. We continue with the example and seek the probability that a student who is known to have forgotten his/her lunch ticket is from pick-up region 2. That is, we seek $p(B_2|A)$. In the following calculations, we use Definition 5.6 twice and Theorem 5.8 once to obtain,

$$p(B_2|A) = \frac{p(B_2 \cap A)}{p(A)} = \frac{p(A \cap B_2)}{p(A)}$$

$$= \frac{p(A|B_2)p(B_2)}{p(A)} = \frac{(.20)(.35)}{0.12} = \frac{7}{12}$$

which is approximately .583.

> **5.9 Theorem (Bayes' Formula):** Let B_1, B_2, ... be events which are mutually exclusive and exhaustive; that is, every sample point belongs to one and only one of the B_i's. Also let A be an event with positive probability, that is $p(A) \neq 0$. Then, for any B_j,
>
> $$p(B_j|A) = \frac{p(A|B_j)p(B_j)}{p(A)} = \frac{p(A|B_j)p(B_j)}{\sum_i p(A|B_i)p(B_i)}.$$

5.8 INDEPENDENCE OF EVENTS

In the example of Figure 5.2, above, $p(A|B) = 2/3$ is not equal to $p(A) = 1/2$. In general, the knowledge whether B occurred could change the odds of the occurrence of A. However, if the occurrence of B does not affect the probability that A occurs, A may be regarded as **independent** of B. Such is the case when two dice are rolled: the face on one die is independent of the face on the other. (See Example 2 of Section 5.6.) Similarly, the outcome of the first toss of a coin does not affect that of the second. This motivates the following.

> **5.10 Definition:** Let A and B be two events in the sample space S. A and B are **independent** (also, **statistically independent**) iff
>
> $$p(A|B) = p(A).$$

An equivalent form to the definition may be obtained with the use of Definition 5.6:

5.11 Theorem: The events A and B are independent if and only if

$$p(A \cap B) = p(A)p(B).$$

Care should be taken not to confuse "independent" with "mutually exclusive." As an example, consider the outcomes of the tossing of two coins. The sample space (with H denoting "heads" and T denoting "tails") is $\{(H,H), (H,T), (T,H), (T,T)\}$. Let A be the event "tails on the first coin" and B the event "tails on the second coin." With $p(H) = p(T) = 1/2$, we have

$$p(A) = p(\{(T,T),(T,H)\}) = \frac{1}{2},$$

$$p(B) = p(\{(T,T),(H,T)\}) = \frac{1}{2},$$

$$p(A \cap B) = p(\{(T,T)\}) = \frac{1}{4} = p(A)p(B).$$

Hence, A and B are independent, *by definition.* Yet (T,T) is a mutual sample point of A and B and so they are not mutually exclusive. (It is of interest to note that, if the probabilities of the sample events are changed, the two events A and B may no longer be independent.)

As another illustration, consider Example 2 of Section 5.6. The event B is "both faces of two dice are odd." If C is the event "the face of the first die is odd" and D is "the face of the second die is odd," $p(C) = p(D) = 3/6 = 1/2$. Since $p(B) = p(C \cap D)$ and since $p(B) = 1/4$, we have $1/4 = p(C \cap D) = p(C)p(D) = (1/2)(1/2)$ and hence C and D are independent events.

The concept of independent events generalizes to more than two events:

5.12 Definition: Let A_1, A_2, $\ldots$, A_n be events in a sample space S.

5.12 Definition—Continued

(i) The A_i's are said to be **pairwise independent** iff, for any $i,j \in \{1, \ldots, n\}$ such that $i \neq j$, A_i and A_j are independent.

(ii) $A_i, \ldots, A_n$ are **independent** iff, for any (two or more) selected $B_1, B_2, \ldots, B_k$ from among the A_i's, so that $2 \leq k \leq n$,

$$p(B_1 \cap B_2 \cap \ldots \cap B_k) = p(B_1)p(B_2) \ldots p(B_k).$$

To illustrate a point of caution, consider the throwing of two dice and the three events defined by: C is "an odd face on the first die," D is "an odd face on the second die" and E is "an even sum of the two faces." It is easy to verify that all individual, as well as conditional, probabilities are equal:

$$p(C) = p(D) = p(E) = \frac{1}{2}$$

and $\quad p(C|D) = p(C|E) = p(D|C) = p(D|E)$

$$= p(E|C) = p(E|D) = \frac{1}{2}.$$

The three events are pairwise independent, as may be seen from the fact that $p(C \cap D) = p(C)p(D)$, etc. However, the sum of two odd faces is always even and thus $C \cap D \cap E = C \cap D$. Consequently,

$$p(C \cap D \cap E) = p(C \cap D) = p(C)p(D)$$

$$= \left(\frac{1}{2}\right)\left(\frac{1}{2}\right) = \frac{1}{4}, \text{ while } p(C)p(D)p(E) = \frac{1}{8}.$$

Thus, the three events, C, D and E, are pairwise independent but not independent.

There are many additional useful results involving independent events. We include here only the following, which is perhaps the most significant.

5.13 Theorem: If A and B are independent events, so are all the combinations of the events and their complements: A and B', A' and B, and A' and B'.

5.9 THE SHAPES OF SAMPLE POINTS AND SAMPLE SPACES

We restrict our attention to cases of random sampling, i.e. to "experiments" in which all sample points are equally likely. (Other cases are not far different from this one; they require that the assigned unequal probabilities either be given or be computable by a given rule.) We consider mostly finite sample spaces, leaving the countably infinite case as an extension.

Let the sample space have n points. Then each such sample point, or rather the simple event whose only member is the sample point, has probability $1/n$. To compute the probability of an event E we need:

1. The number $n = |S|$ of points in the sample space S.

2. The number k of sample points in E—the "favorable" cases. The probability $p(E)$ of E is then k/n.

The following examples illustrate the computation of the sizes of some relatively simple sample spaces.

EXAMPLE 1:
In the case of a single throw of a die, the sample points are single numbers from 1 to 6. $S = \{1,2,3,4,5,6\}$ and $|S| = 6$.

EXAMPLE 2:
In the case of a single throw of two dice, $S = \{(i,j): 1 \leq i,j \leq 6\}$ and $|S| = 36$.

If one die is thrown twice, the sample space is the same as that of a single throw of two dice.

EXAMPLE 3:
A coin is tossed three times. With H and T denoting the outcomes "heads" and "tails," respectively, of one toss of a coin, $S = \{TTT,TTH,THT,THH,HTT,HTH,HHT,HHII\}$ and $|S| = 8$.

EXAMPLE 4:
Tossing a coin until a tail turns up for the first time has the sample points T, HT, HHT, $HHHT$, etc. This is an example of an infinite sample space. If the number of tosses were restricted to, say, 6 and the "failure" sample point of no tail at all included, the sample space would have, in addition to the sample points mentioned, $HHHHT$, $HHHHHT$ and $HHHHHH$. Then $|S| = 7$.

The computation of the total number of sample points in a more complicated sample space, as well as that of the number of favorable cases, is the subject of **combinatorial analysis.** We consider several key cases in the following sections.

5.10 DISTINGUISHABLE EVENTS OF DIFFERENT TYPES—UNORDERED

Here we consider *selection without arrangement,* i.e. the order of the objects selected does not matter.

Consider the set $D = \{0,1, \ldots ,9\}$ of the ten decimal digits and the set $L = \{a,b, \ldots ,z\}$ of the 26 lower case English letters.

Singles: The number of ways to select *one character* that is either from D or from L is the *sum* of the number of ways to select a letter from L and the number of ways to select a digit from D: $26 + 10 = 36$.

Doubles: The number of ways to select a pair, one from D *and* one from L is the product: (10 ways to select a digit) times (26 ways to select a letter) = 260 ways to select one digit and one letter. (A complete listing may be arranged in a rectangular array similar to the multiplication table.)

> *5.14 Theorem:* Let A be a set of m distinguishable objects and let B be a set of n distinguishable objects which are distinct from the members of A.
>
> *(i)* (**The Law of Sum**) There are $m + n$ ways to select an object from *either A or B.*
>
> *(ii)* (**The Law of Pairs**) There are $(m)(n)$ ways to select one object from A *and* one object from B.

Multiples: The selection of one object from each of a number of sets is governed by the following.

> *5.15 Theorem* (**The Law of Product**): Let $A_1, A_2, \ldots, A_k$ be sets of $r_1, r_2, \ldots, r_k$ distinguishable objects, respectively, so that the objects of each set are also distinguishable from the objects of each of the other sets. Then there are $(r_1)(r_2) \ldots (r_k)$ ways of forming a k-tuple by selecting one object from each set.

*5.11 PERMUTATIONS—ARRANGEMENTS OF DISTINGUISHABLE OBJECTS

One Object can be "arranged" in only one way: *a*.

Two distinguishable objects, a and *b*, can be arranged in two ways: the first position can be occupied in either of two ways, *a*-- or *b*--. Once that is decided, there is no choice for occupying the second position, since in each case there is only one candidate left. The resulting two arrangements are then, *ab* and *ba*.

When there are *three distinguishable objects* to arrange, the first position may be occupied in any of three ways, *a*--, *b*-- or *c*--. For each of these ways of occupying the first position, there are only two objects left with which to occupy the second position. Thus, there are (3)(2) = 6 ways in all to occupy the first and second position. (The last position must always be occupied by the remaining object and hence there is no choice left for it—there is only one way to accomplish that.)

For the general case, it is convenient to have the following.

5.16 Definition: For any positive integer *n*, we define *n* **factorial** as $n! = n(n - 1)(n - 2) \ldots (2)(1)$. Also, 0 **factorial** is defined as $0! = 1$.

When there are *n distinguishable objects* to arrange, for any positive integer *n*, there are *n* ways to choose the occupant of the first position. For each of those, there are $n-1$ ways to occupy the second position, etc. The total number of arrangements of *n* distinguishable objects, then is

$$n(n - 1)(n - 2) \ldots (2)(1) = n!$$

If only $r \leqq n$ of the *n* objects are to be selected and arranged, then only the first *r* positions are to be occupied. The number of ways to accomplish that is then

$$n(n - 1) \ldots (n - r + 1)$$

$$= \frac{[n(n - 1) \ldots (n - r + 1)][(n - r) \ldots (2)(1)]}{(n - r) \ldots (2)(1)} = \frac{n!}{(n - r)!}$$

---

*For a discussion of permutation groups, see Section 12.8.

5.17 Definition: A **permutation** of *r* objects is their
ordered linear array. A **permutation of *r* out of *n* objects** is
the ordered linear array of the *r* objects selected from among
n objects.

Where $r \leq n$, $P(n,r)$ denotes the **number of possible
permutations of any *r* out of *n* distinguishable objects.**[1] The
number of permutations of *n* objects is denoted by $P(n,n)$.

A summary of the discussion preceding the definition
appears in the following.

5.18 Theorem: For $r \leq n$, $P(n,r) = \dfrac{n!}{(n-r)!}$

Also, $P(n,n) = n!$

5.12 SAMPLING WITH AND WITHOUT REPLACEMENT

Suppose a four-letter word is to be constructed by choosing
any letter for each of the four positions. Then, there are 26
ways to select a letter for each position. Hence, the Law of
Product of Theorem 5.15 indicates that there are
$(26)(26)(26)(26) = 26^4$ ways of constructing such a word.
(Since each of the four positions is distinguishable from all the
others, the four sets of letters are mutually distinguishable.)

It is important to note that, after selecting a letter for each
position, that letter is *not* removed from the set of letters so that
all 26 letters are available for selection for the next position. In
other words, the letter is **replaced** back in the set. We call such
a process **sampling with replacement.**

5.19 Theorem: The number of ways to select a sample of
size *r* from a set of size *n* ($r \leq n$) with replacement is n^r.

If the four-letter word is to be selected **without replacement,** we may imagine 26 plastic letters in a basket and the
word formed by selecting, one at a time, four plastic letters to
be put side by side. There are 26 ways to select the first letter.
Once that happened, there are only 25 letters left in the basket

---

[1]This notation for the number of permutations is not universal. Another fairly
common notation is $(n)_r$.

and hence only 25 ways to select the second letter. For similar reasons, there are 24 ways to select the third letter and 23 ways to select the fourth.

The number of ways to form the word is then $P(26,4)$, since we are seeking the number of permutations of 4 out of 26 letters. By the Law of Product of Theorem 5.15, this number is $(26)(25)(24)(23) = 358,800$.

It may be of importance to note that, if we occupy the third position first and then the first, fourth and second in order, the number of possible outcomes does not change.

> *5.20 Theorem:*   The number of ways to select a sample of size r from a set of size n ($r \leq n$) without replacement is
>
> $$P(n,r) = \frac{n!}{(n-r)!}$$

5.13 COMBINATIONS—UNORDERED SELECTION WITHOUT REPLACEMENT

We continue with the example of the preceding section. Let four plastic letters be picked out of the basket of 26 such distinct letters, but they are not to be arranged into a word yet.

Consider now the four letters just selected; call them a, b, c and d. The number of different permutations of these four letters is $P(4,4) = 4! = 24$. But for our purposes, *abcd* and *cadb* are the result of the *same unordered* selection, even though they are different permutations. They are *not* to be distinguished here. The number $P(26,4)$ of *ordered* selections of four letters out of 26 is then 24 times bigger than that for the *unordered* selection. Hence, the number we want is

$$\frac{P(26,4)}{P(4,4)} = \frac{358,800}{24} = 14,950.$$

This is the number of **combinations** of 4 letters out of 26. It is called the **binomial coefficient** of 26 over 4 (or 26 *and* 4).

> *5.21 Definition:*   A **combination** of r objects is their unordered array. A **combination of r out of n objects** is an unordered array of the r objects selected from among n objects.
>
> $C(n,r)$ denotes the **number of possible combinations of r out of n distinguishable objects.**

5.22 Theorem: For $r \leq n$,

$$C(n,r) = \frac{P(n,r)}{P(r,r)} = \frac{n!}{(n-r)!r!}.$$

5.23 Definition: The **binomial coefficient of n and** (alternatively, **over**) r is defined as

$\binom{n}{r}$

$$\binom{n}{r} = C(n,r) = \frac{n!}{(n-r)!r!}.$$

The following are useful facts for manipulating expressions which include the binomial coefficient. (The binomial coefficient, as its name may suggest, appears in other contexts, as well.)

5.24 Theorem: Let $r \leq n$. Then,

(i) $\binom{n}{0} = \binom{n}{n} = 1.$

(ii) $\binom{n}{r} = \binom{n}{n-r}.$

(iii) $\binom{n}{r-1} + \binom{n}{r} = \binom{n+1}{r}.$

(iv) $\binom{n}{0} + \binom{n}{1} + \ldots + \binom{n}{n} = 2^n.$

The following two examples illustrate the computations of probabilities involving the binomial coefficient.

EXAMPLE 1:

Selecting a hand of 13 cards at random from a full deck of 52 playing cards can be done in $\binom{52}{13}$ ways. Only one of these combinations has all Clubs. Similarly, there is only one combination for each of the other suits. Thus, the probability of getting a hand all of the same suit is $4 / \binom{52}{13}.$

EXAMPLE 2:

What is the probability that, out of 5 tosses of a coin, there are exactly 3 heads? The number of ways to get 3 heads out of 5 tosses is $\binom{5}{3}$. The total number of possible tosses is 2^5. Thus, the probability we seek is $\binom{5}{3}/2^5 = 10/32 = 5/16.$

5.14 OCCUPANCY PROBLEMS

Many combinatorial problems may be reduced to the random distribution of r balls in n cells. Each ball can be placed in each of n cells and so the number of different distributions is n^r. If the balls are not distinguishable, the number of distinguishable distributions is smaller.

Two balls in two cells could cluster in the following ways: (2,0), (1,1), (0,2). Three balls in two cells: (3,0), (2,1), (1,2), (0,3).

The six faces of a die may be regarded as six cells. If the dice are indistinguishable, the event of 6 dice coming up all different is like having one ball in each of six cells: the 6! = 720 ways this can happen are indistinguishable from each other.

The subject of Section 15 is the solution of occupancy problems.

5.15 UNORDERED SELECTION WITH REPLACEMENT

When seven digits are selected, with possible repetition, the result might look like 1,1,1,3,3,5,6. (The nondecreasing order is accidental. The order does not matter—only the fact of multiple copies of integers does.) This may be accomplished by *copying,* but not removing, the selected digit from the original set of ten digits, and then repeating the process of selection enough times.

We now translate the collection 1,1,1,3,3,5,6 into the **occupancy pattern** of ten bins, each for one of the integers (the order is not important), separated by vertical bars.

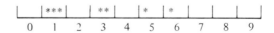

Each digit selected causes the deposit of an * into the corresponding bin. With seven *'s (seven digits to be selected) and 9 bars (the extreme bars are fixed and not to be rearranged), a particular selection of seven digits results in a particular assignment of *'s to seven of the $7 + 9 = 16$ locations. Once that is accomplished, there is no choice left in the positions of the internal bars—they simply go into the remaining 9 positions.

For example, $|**|||*||*|***||||$ represents $0,0,3,5,6,6,6$ and is accomplished by assigning *'s to the 1st, 2nd, 6th, 9th, 11th, 12th and 13th positions, ignoring the first and last bars.

There are $\dbinom{10 + 7 - 1}{7} = \dbinom{16}{7} = C(16,7) = \dfrac{16!}{9!7!} = 11{,}440$

ways of selecting seven digits with replacement.

5.25 Theorem: Where $r \leq n$, the number of ways to select r objects from among n distinct objects, with repetition permitted, is

$$\binom{n + r - 1}{r} = C(n + r - 1, r).$$

EXAMPLE:
In how many ways can four coins be selected out of a large number of pennies, nickels, dimes and quarters?

Selection with replacement of four coins out of four types can be done in $\dbinom{4 + 4 - 1}{4} = \dbinom{7}{4} = \dfrac{(7)(6)(5)}{(1)(2)(3)} = 35$ ways.

(In fact, there are only 34 different amounts of money resulting, since there are two different (distinguishable) combinations which total 40¢; all others are unique.)

5.16 THE HYPERGEOMETRIC DISTRIBUTION

A common type of combinatorial problem is illustrated by the following. There are m white and n black poker chips. From these $m + n$ chips, r are selected at random without replacement, with the order disregarded. What is the probability of having t white chips among the r selected, where $0 \leq t \leq r \leq m$?

The white chips can be chosen in $\dbinom{m}{t}$ ways and the remaining $n - t$ black ones in $\dbinom{n}{r - t}$ ways. Altogether, there are $\dbinom{m + n}{r}$ ways of selecting r out of the $m + n$ chips. Thus the probability we seek is

$$\frac{\dbinom{m}{t}\dbinom{n}{r-t}}{\dbinom{m+n}{r}}.$$

EXAMPLE:

The probability that a hand of 13 bridge cards has a $4 - 3 - 3 - 3$ distribution may be calculated as follows. The probability that there are 4 spades, 3 hearts, 3 diamonds and 3 clubs is

$$\frac{\binom{13}{4}\binom{13}{3}\binom{13}{3}\binom{13}{3}}{\binom{52}{13}}.$$

The same is true if the 4-card suit is hearts, diamonds or clubs. Thus, the probability of a $4 - 3 - 3 - 3$ distribution is

$$4 \times \frac{\binom{13}{4}\binom{13}{3}\binom{13}{3}\binom{13}{3}}{\binom{52}{13}},$$

which is close to 0.11.

6

Graphs

PART I: UNDIRECTED GRAPHS

6.1 ALL KINDS OF GRAPHS

The diagram in each of the parts of Figure 6.1 is commonly called a **graph,** although strictly speaking it is not. The circles, with the labels inside or near them (and even with no labels at all) are called **vertices** (the singular is **vertex**) (also **nodes, points, junctions**) and the connecting line segments are called **edges** (also **lines, arcs, branches**). The edges need not be straight.

Formally, each part of Figure 6.1 is not a graph but the **diagram** of a graph, a visual rendition of a graph. The graph itself is an abstract structure (V,E), where V is a nonempty set (of vertices) and E is a set (of edges) of unordered pairs of

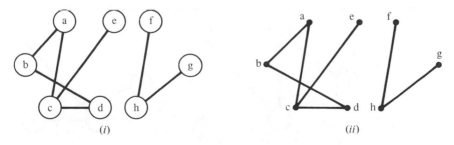

(i)   *(ii)*

Two forms of a graph G

FIGURE 6.1

members of *V*. In the graph pictured in Figure 6.1, the vertices are *a*, *b*, *c*, *d*, *e*, *f*, *g*, *h* and the edges are the *unordered* pairs {*a,b*}, {*a,c*}, {*b,d*}, {*c,d*}, {*c,e*}, {*f,h*}, {*g,h*}.

The visual representation, however, is often far more helpful than the abstract description and thus we abandon the distinction between the two: we adopt the common convention of speaking of a graph and referring to its diagram.

The graph of Figure 6.1 may be regarded as describing acquaintances among people attending a social event. Mr. *b* wishes to be introduced to Miss *e*, whom he does not know. A **path** must be found which **connects** *b* and *e*, using successive edges with common vertices: *b* knows *a*, who knows *c*, who knows *e* and the introduction can take place. This is depicted by the **subgraph** *H* of *G*, shown in Figure 6.2(i). It consists of some of the edges of *G* and all the vertices **incident** to them.

H is a **connected** subgraph, since a path exists between any two of its vertices. Since such a path exists between *b* and *e*, they are **connected vertices**. The path {*b,a*}, {*a,c*}, {*c,e*}, or more simply *b*, *a*, *c*, *e*, is of **length** 3 because it uses three edges to connect *b* and *e*.

Another possible path between *b* and *e* is the chain of edges {*b,d*}, {*c,d*}, {*c,e*}. This is the connected subgraph *F*, shown in Figure 6.2(ii).

Each of the paths *b*, *a*, *c*, *e* and *b*, *d*, *c*, *e* is an **open path,** since its first and last vertices are different. The path *b*, *d*, *c*, *a*, *b* in the original graph *G* (Figure 6.1) is a **loop** (also a **circuit**), because its first and last vertices are the same. It also is a **cycle,** because no other vertex is repeated.

The part of the graph *G*, which consists of the vertices *a*, *b*, *c*, *d* and *e* and of all the edges of *G* which involve these vertices, is also a connected subgraph of *G*; however, it is a **maximal connected** subgraph, because if any other part of *G* is added to

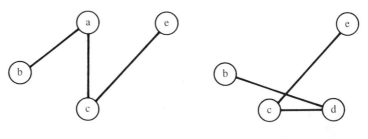

(*i*) Subgraph H of G   (*ii*) Subgraph F of G

FIGURE 6.2

it, the result is no longer connected. Such a subgraph is called a **block,** or a **component,** of *G*. The rest of *G* forms another block: the vertices are *f, g* and *h* and the edges are {*f,h*} and {*g,h*}. When a graph has more than one block, as *G* does, it is **not connected,** or **unconnected.** (Note that "disconnected" is incorrect.)

A characteristic of unconnected graphs is the existence of two vertices with no path between them, that is **unconnected vertices.** Had Mr. "*b*" of the graph of Figure 6.1 wanted an introduction to Miss "*g*," he would be out of luck since *b* and *g* are in two different blocks of *G* and hence they are not connected—there exists no chain of acquaintances to effect the introduction.

In the graphs discussed thus far, there was no direction designated for any edge: the edge {*a,b*} could be written as {*b,a*} and could still be thought of as "*a* and *b* know each other." On the other hand, if such an edge were to describe "*a* is a parent of *b*," it could not be reversed without changing the relationship. In such a case, the edge has a **direction** and this fact is denoted by the addition of an arrowhead oriented properly. If each edge of the graph represents a parent-child relationship, the graph might appear as in Figure 6.3. Such a graph is called a **directed graph** (also **digraph**), since each edge in it is assigned a direction. By contrast, the graph of Figure 6.1 is an **undirected graph.** (The term "graph" often applies to both directed and undirected graphs. The full name "directed graph," or its abbreviation "digraph," should be used for this special case.)

If the original graph of Figure 6.1 were to depict the traffic pattern in the streets connecting certain intersections in the city, one-way streets could be marked with arrowheads show-

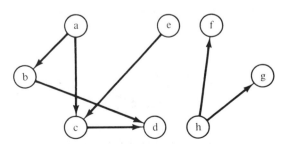

Ancestry graph of G—a directed graph

FIGURE 6.3

ing the direction of traffic. Figure 6.4(i) shows how one component of the graph G may look after the change. The link connecting c and e has two-way traffic, while to reach from b to a one must go through c because of the three one-way streets. Such a graph, some of whose edges are directed and others are undirected, is called a **mixed graph.**

Both an undirected graph and a mixed graph can be converted into a directed graph by inserting two directed edges, one in each direction, for every undirected edge of the original graph. This was done in Figure 6.4 in obtaining part (ii) from part (i).

The use of the term "graph" in the general sense includes graphs with multiple edges connecting two vertices (called **multigraphs**), graphs with edges connecting a vertex with itself (called **pseudographs**), graphs with their edges labeled (called **labeled graphs**), as well as directed graphs, mixed graphs, undirected graphs and other varieties.

The multigraph in Figure 6.5(i) is the graphical essence of the celebrated **Königsberg bridge problem,** to which we return in Section 6.4, below. There are two bridges connecting A and B, and two connecting B and C, a fact represented by two edges connecting each pair of the corresponding vertices.

The pseudograph of Figure 6.5(ii) may be regarded as a chart of who played cards with whom on a certain day among the four people A, B, C and D. (B did not play solitaire and D played nothing but solitaire.)

The strict sense of the term "graph" means an undirected graph with no more than one edge connecting any two vertices (no cycles of length 2) and with no edge connecting a vertex with itself (no cycles of length 1).

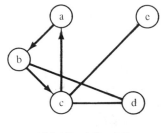   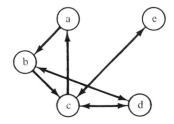

(*i*) A Mixed Graph D   (*ii*) D Converted to a Digraph

FIGURE 6.4

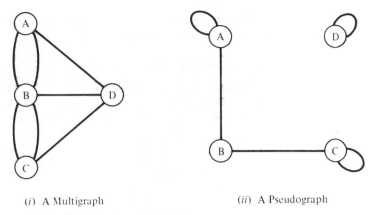

(*i*) A Multigraph   (*ii*) A Pseudograph

FIGURE 6.5

6.2 DEFINITIONS

We use the following sequence of definitions to consolidate and complete the terminology and concepts of the preceding section. The first definition includes basic terminology and concepts. It should be carefully noted that it permits a graph to be without edges (Figure 6.6(i), below) but not without vertices.

6.1 Definition: A **graph** G is a pair $G = (V,E)$, where V is a nonempty set of **vertices** and E is a set of unordered pairs of vertices (of the form $\{v,u\}$, where $v,u \in V$). The pairs of vertices in E are called **edges**, E is called the **edge set** of G and V is called the **vertex set** of G.

G is **finite** iff V and E are finite; otherwise, G is an **infinite** graph. (Unless otherwise stated, we assume all graphs in this book to be finite.)

An edge $\{v,u\}$ is **incident** with the vertices v and u in it (i.e., with the vertices it "joins"). Two vertices v and u are **adjacent** iff the edge $\{v,u\}$ is in E (i.e., iff there is an edge which joins them).

A graph is **directed** iff directions are assigned to the edges; i.e., iff the unordered pairs $\{v,u\}$ in E become ordered pairs (v,u) by specifying for each edge $\{v,u\}$ in E an **initial** (first) vertex v and a **terminal** (second, or last) vertex u. A directed graph is also called by its abbreviated name **digraph.**

A graph G is **undirected** iff directions are assigned to no edges in E. A graph G is **mixed** iff directions are assigned to some edges in E but not to all. ∎

The next definition extends the concept of a graph to include more than one edge connecting two vertices.

6.2 Definition: Let $G = (V,E)$ be a graph. Assign to each pair v and u of vertices from V a nonnegative integer $m(v,u)$, called the **multiplicity** of the pair, according to the following scheme: if $\{v,u\} \in E$ (i.e., there is such an edge in G), then $m(v,u) = 1$ and v and u are called **adjacent vertices;** if $\{v,u\} \notin E$ (i.e., there is no such edge in G), then $m(v,u) = 0$ and v and u are **not adjacent.** G is called a **pseudograph** iff it has at least one edge with identical vertices, i.e. iff there is $v \in V$ with $m(v,v) = 1$. (Such an edge connects the vertex v with itself.)

We extend the definition of a graph to allow more than one edge to join two vertices. The **multiplicity** $m(v,u)$ of a pair of vertices v and u is then the number of edges joining v and u. G is a **multigraph** iff it has at least one pair of vertices whose multiplicity is greater than 1. ∎

The next definition deals with the rudiments of connectedness in graphs.

6.3 Definition: Let $G = (V,E)$ be a graph. A graph $H = (U,F)$ is a **subgraph** of G iff $U \subseteq V$ and $F \subseteq E$.[1] H is a **proper subgraph** of G iff H is a subgraph of G and of at least two vertices and $F \neq E$ or $U \neq V$.

A **path** in an undirected graph[2] is a sequence of edges e_1, $e_2, \ldots, e_n$ from E in which any two successive edges have a common vertex (not shared by the immediately preceding edge); i.e., $\forall i \in \{1,2,\ldots,n-1\}$ there exist v_1, v_2, $v_3 \in V$ such that $e_i = \{v_1,v_2\}$ and $e_{i+1} = \{v_2,v_3\}$. The **length** of the path is the number n of edges in it. A **path between** two vertices v and u is a path $e_1, e_2, \ldots, e_n$ such that v is a vertex in e_1 and u is a vertex in e_n; in such a case, the path is said to **connect** v and u. Two vertices v and u are **connected** iff there exists a path which connects them.

A graph G is **connected** iff every pair of distinct vertices in it is connected; otherwise, G is **unconnected** (also **not connected**). A maximal connected subgraph of G is called a **block** of G (also a **component** and a **connected component** of G). ∎

In the following definition, directions are imposed on the edges of an undirected graph in order to simplify and make precise the descriptions of certain substructures; these

---

[1] Note that H is a graph and therefore U must contain all vertices with which edges in F are incident.

[2] The concept of a path is more natural to directed graphs (Section 6.10).

substructures are used in undirected graphs, even though they are more natural in the context of directed ones.

> **6.4 Definition:** Let $G = (V,E)$ be a graph and let $p = e_1, e_2, \ldots, e_k$ be a path in G. Order[1] each pair of vertices in the path p so that, for any $i \in \{1, 2, \ldots, k-1\}$, the terminal vertex of e_i is identical to the initial vertex of e_{i+1}; i.e., for some $v_1, v_2, v_3 \in V$, $e_i = (v_1, v_2)$ and $e_{i+1} = (v_2, v_3)$. We denote the resulting path p either by its sequence of edges $p = (v_1, v_2), (v_2, v_3), \ldots, (v_{k-1}, v_k)$ or by its sequence of vertices $p = v_1, v_2, \ldots, v_k$.
>
> The path p is **open** iff $v_1 \neq v_k$; it is **simple** iff no edge is repeated; it is **elementary** iff no vertex is repeated. The path p is a **loop** (also a **circuit**) iff $v_1 = v_k$; that is, the path connects a vertex to itself. A loop is **simple** iff it has no repeating edge. A loop is a **cycle** (also an **elementary loop**) iff the only repeated vertex is $v_1 = v_k$. ∎

6.3 THE DEGREE OF A VERTEX

An **isolated** vertex is one which is connected to no vertex (not even to itself). Thus, an isolated vertex is in no edge of the graph. A **null graph** is one with no edges, and thus all its vertices are isolated. The null graph with five vertices is shown in Figure 6.6(i).

At the other extreme is the graph in which any two distinct vertices are adjacent. Such a graph is called **complete** (also

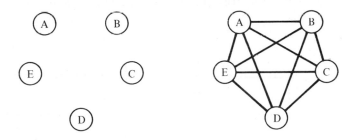

(*i*) The Null Graph with 5 Vertices   (*ii*) The Complete Graph with 5 Vertices

FIGURE 6.6

---

[1]This is always possible by repeating an edge, but reversing its direction, when necessary. The resulting sequence of edges is longer then, but the representation is faithful to the traversal of the path.

universal). An example is shown in Figure 6.6(ii). In it, every vertex is adjacent to each of the remaining four vertices, and thus there are four edges incident with each vertex. This number of edges incident with a vertex is called the **degree** of the vertex. It means the same for multigraphs and pseudo-graphs.

Take, for example, Figure 6.7 of the next section. In it, the degree of A is $\deg(A) = 3$, $\deg(B) = 5$, $\deg(C) = 3$ and $\deg(D) = 3$. These numbers are significant in the next section.

A graph in which every vertex has the same degree is called **regular**. The graph in Figure 6.6(ii) is therefore regular, as any complete graph must be.

> **6.5 Definition:** Let $G = (V,E)$ be a graph and let v be a vertex of G (i.e., $v \in V$). The **degree** of v is denoted by $\deg(v)$ and is defined as the number of distinct edges incident with v. A graph is **regular** iff every one of its vertices has the same degree.

6.4 EULER PATH—THE KÖNIGSBERG BRIDGE PROBLEM

The **Königsberg bridge problem** was solved by the Swiss mathematician Leonhard Euler (pronounced Oiler) around 1736 with the aid of the degrees of vertices. The problem was to traverse exactly once each of the seven bridges on the river Pregel in the Prussian town Königsberg. A map of the bridges

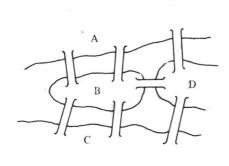

(i) The Königsberg Bridges

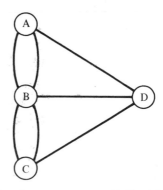

(ii) The Corresponding Multipraph

FIGURE 6.7

appears in Figure 6.7(i); the multigraph essence of it is shown in Figure 6.7(ii). The kind of path required for the traversal of the bridges has since been called an **Euler path,** as described in the following definition.

> *6.6 Definition:*   Let $G = (V,E)$ be a graph. An **Euler path** in G is a path which traverses each edge of G exactly once. An **Euler circuit** (or **Euler loop**) is a closed Euler path, i.e. an Euler path whose first and last vertices are identical. An **Euler graph** is a graph which possesses an Euler circuit.

The following sequence of relatively simple results leads to the negative solution of the Königsberg bridge problem—it cannot be done.

> *6.7 Theorem:*   The sum of the degrees of all the vertices of a graph (alt., multigraph) is even. If the graph (alt., multigraph) has *n* edges, then the sum of the degrees of the vertices of the graph is 2*n*.

As an immediate consequence of Theorem 6.7, we have,

> *6.8 Theorem:*   The number of vertices of odd degree in any graph (alt., multigraph) is even.

Since an intermediate vertex in an Euler path must be exited as many times as it is entered, we have,

> *6.9 Theorem:*   Any vertex in an Euler path, with the possible exception of the first and the last, must be of even degree.

In the multigraph of Figure 6.7, all vertices are of odd degree. As a consequence, it does not possess an Euler path. Thus, the Königsberg bridges cannot be traversed as required.

> *6.10 Theorem:*
>
> *(a)* A graph possesses an Euler path iff it is connected and has zero or two vertices of odd degree.
> *(b)* A graph is an Euler graph (possesses an Euler circuit) iff it has no vertices of odd degree.

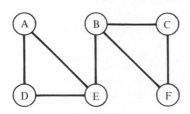

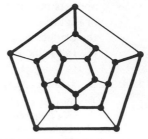

(*i*) A Graph with a Hamiltonian Path
       but no Hamiltonian Cycle

(*ii*) A Hamiltonian Cycle through
        Hamilton's Traveller's Dodecahedron

FIGURE 6.8

6.5 HAMILTONIAN PATH—THE TRAVELING SALESMAN PROBLEM

The traveling salesman problem is easy to state but, apparently, difficult to solve—no efficient solution exists. A salesman must visit each of a number of cities and return to his starting point. How can he do so covering the smallest possible distance?

This is a generalization of a somewhat simpler problem, and neither of them has an efficient solution as yet. The simpler problem is to find a cycle which passes each vertex of a graph exactly once (returning to the point of departure). Such a cycle is called a **Hamilton cycle,** or a **Hamiltonian cycle,** after the

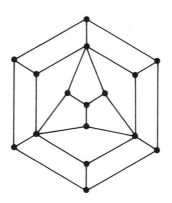

A graph with no Hamiltonian path

FIGURE 6.9

Irish mathematician Sir William Hamilton (1805–1865). A graph which possesses such a cycle is a **Hamiltonian graph.**

A **Hamiltonian path** is a path which passes through every vertex of a graph exactly once. A graph may possess a Hamiltonian path without possessing a Hamiltonian cycle, as may be seen in Figure 6.8(i).

A graph which has no Hamiltonian path is shown in Figure 6.9. Puzzling out why it is so may prove interesting to the reader.

6.6 PLANAR GRAPHS; EULER'S FORMULA

The designer of printed circuits must not let connections (leads) intersect unintentionally, lest unwanted paths ruin the circuit. The designers of highway systems wish to avoid unnecessary crossings of limited-access highways, since the added vertical levels are costly. Many other examples exist where the edges of a graph in the plane should meet only at the vertices they share.

> *6.11 Definition:*   A graph is **planar** iff it can be drawn (in the *plane*) so that its edges intersect only at their vertices.

The graph in Figure 6.10 is planar, because it may be drawn as in part (ii).

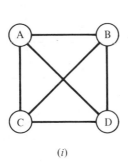

(i)

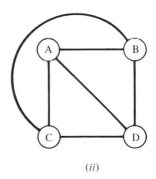

(ii)

Two renditions of a planar graph

FIGURE 6.10

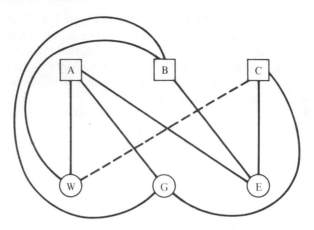

The utility graph

FIGURE 6.11

An example of a graph which is *not* planar is the **utility graph** of Figure 6.11. It depicts the connections of three houses *A*, *B* and *C* with the utilities *W* (water), *G* (gas) and *E* (electricity). In the figure, houses *A* and *B* are connected to all three utilities and house *C* is connected to *E* and *G* and the connections do not cross each other. But, there is no way to connect *C* to *W* without digging across another existing connection. In fact, there exists no arrangement which permits all nine connections without unwanted intersections.

Another graph which is not planar is the complete graph with five vertices, shown in Figure 6.6(ii), above. It is also called the **star graph.** It is shown again in Figure 6.12(ii), along with the customary form of the utility graph (Figure 6.12(i)).

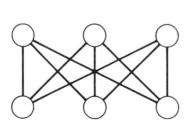

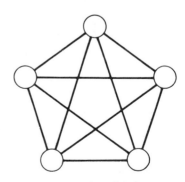

(*i*) The Utility Graph   (*ii*) The Star Graph

FIGURE 6.12

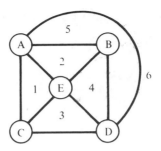

The regions of a planar graph

FIGURE 6.13

The facts governing the planarity of graphs make use of two concepts not covered above. A **region** of a planar graph is an area of the plane bounded by edges and which contains no edges or vertices. The region "outside" the graph is the only infinite region of the graph.

For example, the planar graph of Figure 6.13 has six regions, as numbered. Region 6 is the infinite one. If we denote the number of vertices of the graph by v, the number of edges by e and the number of regions by r, we have here $v = 5$, $e = 9$ and $r = 6$. The relationship $5 - 9 + 6 = 2$ holds for all planar graphs, as is stated in the following.

> **6.12 Theorem (Euler's Formula):** Let G be a connected planar graph with v vertices, e edges and r regions. Then,
>
> $$v - e + r = 2.$$

Euler's formula may be used to show that a graph is not planar, in the event the graph does not satisfy the formula. In fact, regions are defined only for planar graphs. In order to tell whether a graph is planar, another concept is needed, that of contracting a graph. It amounts to "squeezing together" two adjacent vertices and eliminating the edge between them. This process is called a **contraction.**

The graph in part (ii) of Figure 6.14 was obtained from part (i) by the contraction $A, B \rightarrow F$; that is, deleting the edge $\{A,B\}$, removing A and B from the vertex set, inserting the new vertex F into the vertex set, and replacing A by F in the edge $\{A,E\}$ and B by F in the edges $\{B,C\}$ and $\{B,D\}$.

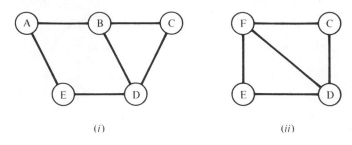

A contraction

FIGURE 6.14

We say that a graph G is **contractible** to a graph H iff H can be obtained from G by a sequence of contractions. It is now possible to state necessary and sufficient conditions for the planarity of a graph. These conditions consist of the absence of (essentially) the two graphs of Figure 6.12.

> **6.13 Theorem (Kuratowski):** A graph is planar iff it has no subgraph which is contractible to either the utility graph or the star graph.

6.7 TREES AND SPANNING TREES

The connected graph of Figure 6.15(i) is not a **tree,** because it has a cycle (A,C,D,A). The graph in part (ii) is a **tree,** since it is connected and it has no cycles. It was obtained from the graph of part (i) by removing one edge. Similarly, the graph in part (iii) is not a tree and the one in (iv) is. The graph in part (v) of Figure 6.15 is not a tree; it possesses cycles. It was necessary to remove three edges in order to "break" all cycles and obtain a tree. Two of the ways which accomplish that are shown in parts (vi) and (vii) of Figure 6.15. In fact, each of the trees in parts (vi) and (vii) is a **spanning tree** of the graph in part (v), since it includes all the vertices of the latter.

> **6.14 Definition:** A connected graph is a **tree** iff it has no cycles. A graph is a **forest** iff every connected component (block) of it is a tree.
> Let $G = (V,E)$ be a connected graph. A subgraph $H = (U,F)$ of G is a **spanning tree** of G iff H is a tree and $U = V$ (and every vertex of G is in an edge of H).

(When one vertex of a tree is designated as a main vertex, as a **root,** the entire tree gains an orientation and becomes a **rooted tree.** Rooted trees and **directed trees** are discussed in Section 6.15, below.)

A connected graph may well have more than one spanning tree. It is possible to obtain a spanning tree of a graph by repeatedly eliminating ("breaking") cycles by removing edges from any existing cycle until no cycles exist. A more construc-

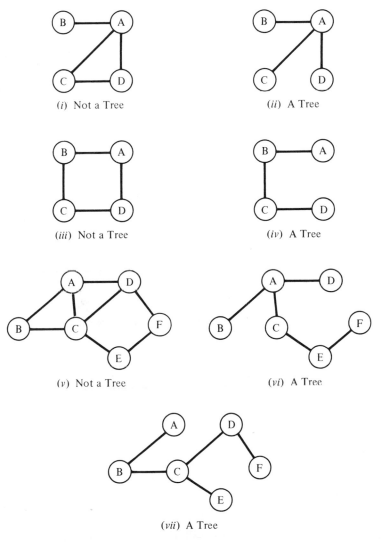

(*i*) Not a Tree

(*ii*) A Tree

(*iii*) Not a Tree

(*iv*) A Tree

(*v*) Not a Tree

(*vi*) A Tree

(*vii*) A Tree

FIGURE 6.15

tive approach is given in Algorithm 6.15. In brief, it starts with all the vertices of the graph G and none of the edges. It then adds one edge at a time from the edges of G, as long as the added edge does not form a cycle.

> ### 6.15 Algorithm (Spanning Tree of a Connected Graph): Given a connected graph $G = (V, E)$, to construct a spanning tree $T = (V, F)$ of G.
>
> **Step 1.**   Initially, F is the empty set of edges. (Do Step 2.)
>
> **Step 2.**   Select an edge e from E. Add e to F if this does not create a cycle in T; i.e., replace F by $F \cup \{e\}$. In any case, delete e from E; i.e., replace E by $E - \{e\}$. (Do Step 3.)
>
> **Step 3.**   If every vertex of G is incident with at least one edge in F, stop; the spanning tree T is complete. If not, do Step 2. ∎

The following several results are easy to verify and are often helpful.

> ### 6.16 Theorem: A tree of v vertices has exactly $v - 1$ edges.

> ### 6.17 Theorem: Inserting an edge $\{v, u\}$ to a tree in which v and u are nonadjacent vertices creates a cycle, and the graph is no longer a tree; it is a graph with exactly one cycle.

A **proper path** between two vertices of a graph is a path in which no edge appears twice.

> ### 6.18 Theorem: In any tree, any two vertices are connected by one and only one proper path.

> ### 6.19 Theorem: A graph is connected iff it contains a spanning tree.

6.8 ECONOMY TREES

A frequent application of spanning trees concerns the finding of a **minimal-cost network**: connect a number of cities with roads in the cheapest way; lay water-main pipes through a number of streets in the cheapest way; from existing railroad links among a number of cities, select connecting links of railroad to repair and maintain in the cheapest way so that each city has railroad service; and the like.

Essentially, the problem is this: Starting with a connected graph whose edges are assigned weights (distances between cities, cost of laying pipe between intersections, etc.), find a connected subgraph with the full vertex set, so that the sum of the weights on the edges of the subgraph is the smallest possible. Such a subgraph is always a spanning tree of the graph and is called an **economy tree.**

The method used in Algorithm 6.20 for finding an economy tree of a connected graph is this: Keep collecting the "cheapest" links (edges), as long as no cycles are formed. If there is more than one cheapest link which causes no cycle, select any one of them.

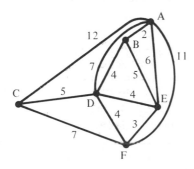

(*i*) The Network

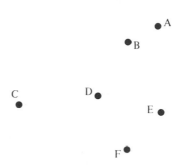

(*ii*) Start with all the Vertices and no Edges

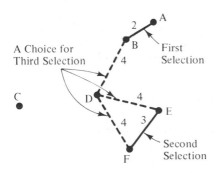

(*iii*) Choose the First Two
Cheapest Links

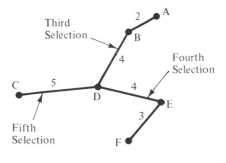

(*iv*) The Finished Graph, Using One Way
of Choosing 3rd and 4th Links

Constructing an economy tree

FIGURE 6.16

Figure 6.16 shows the step-by-step construction of an economy tree for the network in part (i). The weights associated with the edges may be regarded as distances between corresponding cities. Roads are curved because of right-of-way and elevation problems.

The first three parts of the figure are self-explanatory. In part (iii), any of the edges *BD*, *DE* and *DF* may be selected in the third place. If *BD* is selected, either of the remaining two could be selected fourth; if instead *DE* or *DF* is selected third, the other would make a cycle (*DEFD*) if selected fourth and so *DB* would be the only alternative. Part (iv) of Figure 6.16 reflects the choice of *BD* as the third link and *DE* as the fourth.

Now, there are two links of weight 5, but *BE* will form a cycle (*BDEB*). This would still be the case had the third and fourth links been *BD* and *DF*; *BE* would then form the cycle *BDFEB*. Thus *CD* is the fifth link selected and the economy tree is complete.

6.20 Algorithm (Economy Tree of a Connected Weighted Graph): Given a connected graph $G = (V,E)$ with a positive integer $c(e)$, called the **cost** of e, assigned to each edge e from E; to construct an economy tree $T = (V,F)$ for G.

Step 1.   Initially, F is the empty set of edges. (Do Step 2.)

Step 2.   Find the smallest cost c of any edge e in E, and let $R \subseteq E$ be the set of all edges in E with cost $c(e) = c$. (Do Step 3.)

Step 3.   Remove all the edges in R from E; i.e., replace E by $E - R$. (Do Step 4.)

Step 4.   Choose any one edge e from R. Add e to F (i.e., replace F by $F \cup \{e\}$), if and only if this creates no cycle in T. In either case delete e from R; i.e., replace R by $R - \{e\}$. (Do Step 5.)

Step 5.   If each vertex of T is adjacent to at least one vertex (is in at least one edge) of T, stop; T is complete. (Otherwise, do Step 6.)

Step 6.   If R is empty, do Step 2. Otherwise, do Step 4. ∎

Proving that the algorithm always results in an economy tree is not difficult and may be entertaining.

6.9 BIPARTITE GRAPHS AND MATCHING

The set of variables of the utility graph of Figure 6.12(i) (Section 6.6) may be divided into two disjoint parts (the utilities and the houses) so that any edge of the graph connects a vertex of one part with a vertex of the other. No edge connects two vertices in the same part.

The graph in Figure 6.17 is of this sort. The vertex set V may be divided into $U = \{x,y,z,w\}$ and $W = \{C,D,E,F,G\}$. Every edge of the graph connects one vertex from U with one vertex from W. Such a graph is called a **bipartite graph** (meaning, a graph of two parts) and the two parts U and W of V are called **complementary vertex sets.**

It may be difficult to tell of a graph by appearance only if such a separation of the vertex set is possible. The following theorem states when it can be done.

6.21 Theorem: A graph $G = (V,E)$ is bipartite iff every one of its circuits has even length.

An important application of bipartite graphs is that of **matching.** Regard the vertices x, y, z and w in U as persons and C, D, E, F, G of W as jobs they may be trained to perform. The edges $\{z,E\}$ and $\{z,F\}$ indicate that person z can do jobs E and F. To "match U to (or into) W" means here to assign to each person from U a job from W which he can do. The heavy lines in Figure 6.17 show such a matching assignment.

A **matching** (also, **complete matching**) of U to (also into) W in a bipartite graph $G = (V,E)$ with complementary vertex sets U and W is a selection of edges of G so that every vertex from U

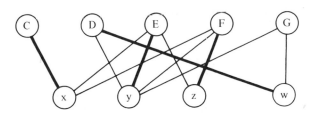

A bipartite graph

FIGURE 6.17

is in a selected edge and no vertex is in two edges.

> **6.22 Theorem:**   Let $G = (V,E)$ be a bipartite graph with complementary vertex sets U and W. Then G has a complete matching of U to W iff, for every positive integer $n \in \{1, \ldots, \#(U)\}$, every n vertices from U are connected to at least n vertices of W.

The condition in Theorem 6.22 (sometimes called the **diversity condition**) can be cumbersome to check in many graphs. There is, however, a criterion, which is only a *sufficient* condition, but which is easier to verify in a bipartite graph and which guarantees a complete matching:

> **6.23 Theorem:**   Let G be a bipartite graph with complementary vertex sets U and W. There exists a complete matching of U to W if there exists a positive integer n such that:
>
> *(i)* every vertex in U is adjacent to at least n vertices in W;
> *(ii)* every vertex in W is adjacent to at most n vertices in U.

Thus, for example, if there are 20 workers on a construction site, there are 20 different jobs to be done, every worker can do at least three of these jobs and every one of the jobs can be done by at most three workers, then all can work at the same time and all jobs will at least be started.

PART II: DIRECTED GRAPHS

6.10 DIRECTED GRAPHS

A graph is **directed** when directions are assigned to the edges. In the undirected graph in Figure 6.18(i), the two vertices A and D of the edge $\{A,D\}$ are not ordered. In the corresponding edge in part (ii) of the figure, the vertex D is **initial** and the vertex A is **terminal;** the direction assigned to the edge is *from D to A*. The edge becomes a **directed edge** and is written as the *ordered* pair (D,A).

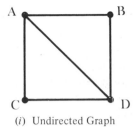

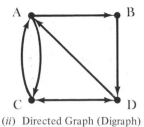

(*i*) Undirected Graph   (*ii*) Directed Graph (Digraph)

FIGURE 6.18

In case both directions are assigned to an edge, as with $\{A,C\}$ in the example, there result two *different* directed edges (A,C) and (C,A). This may appear as two distinct arrows, as in the case of $\{A,C\}$, or as one arrow with both arrowheads, as in the case of $\{C,D\}$; customs vary. A **directed multigraph** is a directed graph with more than one directed edge from a vertex v to a vertex u in at least one case.

The directed graph in Figure 6.18(ii) may be described as $G = (V,E)$, where the **vertex set** is $V = \{A,B,C,D\}$ and the **directed-edge set** is $E = \{(A,B), (A,C), (C,A), (B,D), (D,A), (C,D), (D,C)\}$.

6.24 Definition: A **directed graph** (also **digraph**) is a pair $G = (V,E)$, where V is a nonempty set of **vertices** (V is finite if G is **finite**) and E is a set of **directed edges** (E is finite if G is **finite**) of the form (v,u), where v and u are vertices from V; v is the **initial vertex** and u is the **terminal vertex** of the edge.

The number of edges in which a vertex v is initial is called the **outdegree** of v; the number of edges in which v is terminal is called the **indegree** of v. Both outdegree and indegree of v are called **local degrees** of v.

The terms **subgraph** and **proper subgraph** are as in Definition 6.3: $H = (U,F)$ is a **subgraph** of $G = (V,E)$ iff H is a graph, $U \subseteq V$ and $F \subseteq E$. (Note that the edges in F preserve the directions they are assigned in E.) H is a **proper subgraph** of G iff H is a subgraph of G and of at least two vertices and $F \neq E$ or $U \neq V$.

A **path** (also a **directed path**) in a directed graph $G = (V,E)$ is a sequence of edges $p = (v_0,v_1),(v_1,v_2),\ldots,(v_{n-1},v_n)$ of G in which the terminal vertex of one edge is the initial vertex of the next. A directed path may also be denoted by the sequence of vertices traversed, $p = v_0, v_1, \ldots, v_{n-1}, v_n$. The path p is said to be a path **from** v_0 **to** v_n. The **length** of the path p is the number n of edges in the path. ∎

The following is a simple but important fact about directed graphs.

    6.25 Theorem: If there is a path from a vertex v to a vertex u of a directed graph with n vertices, then there is a path from v to u of length at most $n-1$.

The terms **multigraph, pseudograph, connected vertices, block** or **component, open path, simple** or **elementary path, loop** or **circuit, cycle** or **elementary loop,** retain their definitions from Section 6.2 if the term "graph" is replaced by "directed graph" (and hence a "path" is a directed path). A directed graph is **connected** iff it is connected as an undirected graph; i.e., when directions are ignored.

In many instances, the mere representation of a structure or a problem as a directed graph is enough of an advantage, since then visual tracing is possible in it and since techniques and results of graph theory may be employed. In the remainder of this chapter, several such results and several key examples of applications of directed graphs are exhibited.

We close this section with an example helpful to applications in which every pair of members of a set must be compared and one of them selected, perhaps as the winner. A round-robin tournament is such an illustration.

A **complete directed graph,** also called a **tournament,** is a directed graph which is a complete undirected graph when the directions are ignored, and in which each undirected edge is assigned exactly *one* direction. A five-player tournament is shown in Figure 6.19. In the diagram, an arrow from B to D indicates that B won his match with D.

In any tournament, there is always a directed path through all vertices. If, however, one player never won a match, such a path must end with that player. If a group F of players $(F \subset V)$ was **outclassed** by their opponents (that is, no player in F won a match against any player in $V - F$), it is impossible to find a (directed) circuit (a closed path) through all the vertices. However, if there is no outclassed group, such a circuit exists and no player can be declared the strongest: for any two players v and u, there is a chain of matches in which v beat v', who beat v'', who beat ... who beat u. In Figure 6.19, such a circuit is shown in the heavier directed edges, $AEBDCA$.

We say that, in a complete directed graph $G = (V,E)$, a subset F of E is an **outclassed** subset of vertices iff there exists no edge (v,u) such that $v \in F$ and $u \in E - F$.

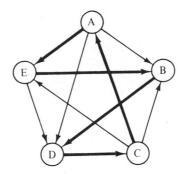

A five-player tournament

FIGURE 6.19

6.26 Theorem: Let $G = (V,E)$ be a complete directed graph.

(i) There is a directed path through all the vertices of G.

(ii) If v is not in an outclassed subset of V, then there is a directed path from v through all the vertices of G.

(iii) If G has no outclassed proper subset of V, then there is a circuit through all the vertices of G.

6.11 THE GRAPH OF A RELATION

Relations on a set are discussed in Chapter 7. As an example of a relation on a set, consider the round-robin tournament at the end of the preceding Section 6.10. The set is

$$P = \{A,B,C,D,E\}$$

of five players and the relation $\mathcal{R}$ on P may be described as "x beat y" for two players x and y from P. The relation $\mathcal{R}$ is then a set of ordered pairs from P.

$$\mathcal{R} = \{(A,B),(A,D),(A,E),(B,D),(C,A),(C,B),(C,E),$$
$$(D,C),(E,B),(E,D)\}.$$

The ordered pairs of the relation may be regarded as the directed edges of a directed graph. In fact, any relation on a set may be represented as a directed graph, by taking the vertex set as the set of the relation and the directed edges as the ordered pairs of the relation.

In our example, the graph of the relation is $G_{\mathcal{R}} = (P, \mathcal{R})$. That is, the vertex set is the set P and the directed-edge set is the *relation* itself; i.e., the related pairs are the directed edges of the graph. The result, of course, is the tournament graph of Figure 6.19, which is repeated in Figure 6.20.

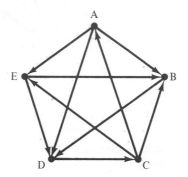

The graph of the relation $\mathcal{R}$ on P

FIGURE 6.20

Special properties of relations are reflected in the graphs of such relations. They are discussed when encountered in Chapter 7.

6.12 THE RELATION OF A DIRECTED GRAPH

Just as a relation on a set may be represented by a directed graph, so can a directed graph be represented by a relation, in the obvious way. If the graph does not depict a particular situation but is given just as a set V of vertices and a set E of directed edges, then the relation acquires no particular meaning beyond the abstract: The relation is $\mathcal{W}$, the set of ordered pairs which in the graph are the directed edges, as is shown in Figure 6.21. The graph in part (i) of the figure depicts the one-way and two-way traffic pattern around the four intersections 1, 2, 3 and 4 in a city; the relatedness of a pair of intersections may be written as $x \, \mathcal{W} \, y$ and described as "from x it is possible to reach y directly."

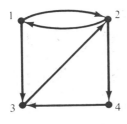

The Set is S = {1, 2, 3, 4}

The Relation W on S is

W = {(1, 2), (1, 3), (2, 1), (2, 4), (3, 2), (4, 3)}

(*i*) A Directed Graph G   (*ii*) The Relation of G

FIGURE 6.21

6.13 THE ADJACENCY MATRIX OF A GRAPH

We continue with the examples of the preceding two sections. The relation $\mathscr{R}$ of the tournament, as well as the graph (repeated in Figure 6.22(i)), may also be represented by an array, called a **matrix** (Figure 6.22(ii)). Although a matrix is rectangular in general, the matrix of a graph is always square.

The rows and columns of the matrix are labeled by the vertices of the directed graph. (When the labels do not show in the label row and the label column, their order must be obvious or must have been established.)

The entry in row A and column B, called the (A,B)-entry, is 1 because there is a directed edge in the graph from vertex A to vertex B. Since no such directed edge exists from B to A, the entry (B,A) in row B and column A is 0.

The matrix in Figure 6.22(ii) shows all adjacencies of the graph and the orientation of each directed edge. For that reason it is called the **adjacency matrix** of the graph.

The adjacency matrix of the graph of Figure 6.21(i) is shown in Figure 6.22 (iv), along with the graph.

The **main diagonal** of a square matrix is the diagonal from top-left to bottom-right, whose entries are (v,v) for each v. If a directed graph has no edge connecting a vertex to itself, all the main-diagonal entries are 0, as is the case in Figure 6.22. However, in a directed pseudograph, where an edge can connect a vertex to itself, entries on the main diagonal may be 1, as is the case in Figure 6.23.

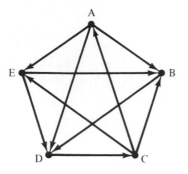

	A	B	C	D	E
A	0	1	0	1	1
B	0	0	0	1	0
C	1	1	0	0	1
D	0	0	1	0	0
E	0	1	0	1	0

(*i*) The Tournament Graph H   (*ii*) The Adjacency Matrix of H

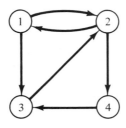

	1	2	3	4
1	0	1	1	0
2	1	0	0	1
3	0	1	0	0
4	0	0	1	0

(*iii*) A Graph G   (*iv*) The Adjacency Matrix of G

FIGURE 6.22

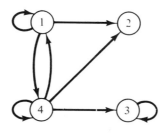

	1	2	3	4
1	1	1	0	1
2	0	0	0	0
3	0	0	1	0
4	1	1	1	1

(*i*)   A Directed Pseudograph P   (*ii*)   The Adjacency Matrix of P

FIGURE 6.23

If more than one directed edge leads from a vertex v to a vertex u, it may be desirable to represent the directed multi-graph by a matrix which shows not only the adjacencies of vertices but also the multiplicity of such connecting edges. (For example, it may be desirable to record the number of messages transmitted from each station to any station in a network.) This may be accomplished by assigning the number of edges from v to u as the (v,u)-entry of the matrix. The resulting matrix may now be called the **multiplicity matrix** of the graph (although this is not a universally used term). An example is shown in Figure 6.24. (It should be noted that, even in a directed multigraph, if only the *fact* of adjacency is of interest, the adjacency matrix suffices.)

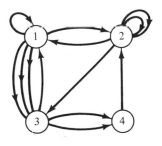

	1	2	3	4
1	1	1	3	0
2	1	2	1	0
3	1	0	0	2
4	0	1	0	0

(*i*) A Directed Multi-pseudograph M   (*ii*) The Multiplicity Matrix of M

FIGURE 6.24

6.14 THE PATH-LENGTH MATRIX OF A GRAPH

The adjacency matrix (and the multiplicity matrix) may be used to compute the number of paths of any given length from a vertex v to a vertex u. It may also be used to compute whether any path from v to u exists. As a consequence, such computation reveals whether the graph is connected. For this purpose, however, we must use multiplication of matrices.

In the following example of **matrix multiplication**, we multiply a 4×3 matrix A by a 3×5 matrix B:

$$A = \begin{bmatrix} 1 & 2 & 0 \\ 2 & -5 & 1 \\ 0 & 0 & 6 \\ 0 & 1 & 0 \end{bmatrix} \qquad B = \begin{bmatrix} 6 & 2 & -1 & 0 & 1 \\ 2 & 0 & 6 & 1 & -3 \\ 1 & 1 & 0 & 0 & 0 \end{bmatrix}$$

The entries in this example are integers, but they could be any real numbers, or elements of any system for which "addition" and "multiplication" is defined.

To find the (2,3)-entry of the product $A \cdot B$ (often refered to as $(A \cdot B)(2,3)$), we isolate *row* 2 of A and *column* 3 of B and then perform the moves illustrated in Figure 6.25.

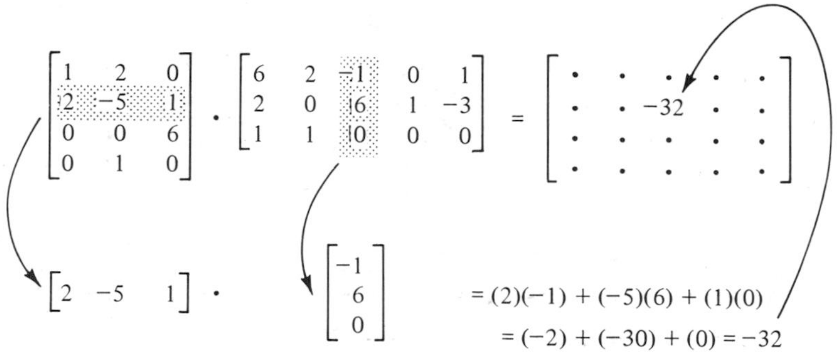

Finding the (2,3)-entry of $A \cdot B$

FIGURE 6.25

The illustration in Figure 6.25 is self-explanatory; it is to be imitated for each of the remaining entries of $A \cdot B$, according to the following instructions: Let row i of A be $[a_1 \ a_2 \ \dots \ a_n]$ and column j of B be

$$\begin{bmatrix} b_1 \\ b_2 \\ \cdot \\ \cdot \\ \cdot \\ b_n \end{bmatrix}.$$

Then the (i,j)-entry of $A \cdot B$ is $\Sigma_{k=1}^{n} (a_k)(b_k) = (a_1)(b_1) + (a_2)(b_2) + \ldots + (a_n)(b_n)$. The reader may verify that in the example of Figure 6.25,

$$A \cdot B = \begin{bmatrix} 10 & 2 & 11 & 2 & -5 \\ 3 & 5 & -32 & -5 & 17 \\ 6 & 6 & 0 & 0 & 0 \\ 2 & 0 & 6 & 1 & -3 \end{bmatrix}.$$

It should be noted that the number of columns of A must equal the number of rows of B. Of course, when both matrices of the product are square of the same size, as in the computation of the **path-length matrices** of a graph, the matching is automatic.

We illustrate with the matrix A of the graph in Figure 6.22, and multiply it by itself.

$$A \cdot A = A^2 = \begin{bmatrix} 0 & 1 & 1 & 0 \\ 1 & 0 & 0 & 1 \\ 0 & 1 & 0 & 0 \\ 0 & 0 & 1 & 0 \end{bmatrix} \cdot \begin{bmatrix} 0 & 1 & 1 & 0 \\ 1 & 0 & 0 & 1 \\ 0 & 1 & 0 & 0 \\ 0 & 0 & 1 & 0 \end{bmatrix} = \begin{bmatrix} 1 & 1 & 0 & 1 \\ 0 & 1 & 2 & 0 \\ 1 & 0 & 0 & 1 \\ 0 & 1 & 0 & 0 \end{bmatrix}$$

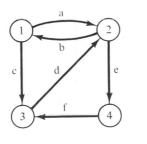

(i) The Graph G of A

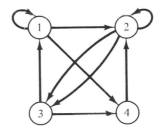

(ii) The Graph of A^2

FIGURE 6.26

The graph of the resulting matrix is shown in Figure 6.26(ii). The edges of the original graph G are labeled in Figure 6.26(i), so that the origins of the edges of the graph of A^2 can be traced. To identify the edge (1,2) from vertex 1 to vertex 2 in the graph of A^2, rewrite row 1 and column 2 of A, replacing the 1's in the matrix A by the names of the corresponding edges of the graph:

$$[0 \quad a \quad c \quad 0] \cdot \begin{bmatrix} a \\ 0 \\ d \\ 0 \end{bmatrix}$$

$$= (0)(0) + (a)(0) + (c)(d) + (0)(0) = (c)(d).$$

It is thus the path cd which is recorded as the edge (1,2) in the graph of A^2.

Similarly, the (1,1) edge in the graph of A^2 is the path ab in G. The entry "2" in row 2 and column 3 of the matrix A^2 signifies the existence of two paths of length 2 from vertex 2 to vertex 3 of G, bc and ef. The entry "0" in row 2 and column 4 of the matrix A^2 means that there exists no path of length 2 in G from vertex 2 to vertex 4.

A^2 is thus the **matrix of path-length** 2 of G.

For any positive integer $n > 1$, A^n is *defined* by $A^n = A^{n-1} \cdot A$, for any square matrix A. We thus compute A^3 in the manner described above:

$$A^3 = A^2 \cdot A = \begin{bmatrix} 1 & 1 & 0 & 1 \\ 0 & 1 & 2 & 0 \\ 1 & 0 & 0 & 1 \\ 0 & 1 & 0 & 0 \end{bmatrix} \cdot \begin{bmatrix} 0 & 1 & 1 & 0 \\ 1 & 0 & 0 & 1 \\ 0 & 1 & 0 & 0 \\ 0 & 0 & 1 & 0 \end{bmatrix}$$

$$= \begin{bmatrix} 1 & 1 & 2 & 1 \\ 1 & 2 & 0 & 1 \\ 0 & 1 & 2 & 0 \\ 1 & 0 & 0 & 1 \end{bmatrix},$$

which shows all paths of length 3 in G.

In the following theorem, we assume the vertices of the graph to be ordered, and that the order is preserved in the adjacency and the multiplicity matrices.

6.27 Theorem: Let G be a directed graph (including multigraph and pseudograph) with vertices $1, 2, \ldots, n$, and let A be its adjacency matrix (multiplicity matrix for multigraphs and pseudographs). Then, for every positive integer k, the (i,j)-entry of A^k is the number of paths of length k from vertex i to vertex j in G.

According to Theorem 6.25 of Section 6.9, above, if the vertex u can be reached from the vertex v at all, it can be so reached in fewer than four steps, since there are only four vertices in the graph of the example. Hence, if there is a path from vertex v to vertex u, it shows in A, in A^2 or in A^3. We perform **matrix addition** on these three matrices and the sum will account for all of the above paths.

Two matrices of equal size are added by adding their corresponding entries, the (i,j)-entry of the first is added to the (i,j)-entry of the second for every i and j. We show the addition in two steps:

$$A + A^2 = \begin{bmatrix} 0 & 1 & 1 & 0 \\ 1 & 0 & 0 & 1 \\ 0 & 1 & 0 & 0 \\ 0 & 0 & 1 & 0 \end{bmatrix} + \begin{bmatrix} 1 & 1 & 0 & 1 \\ 0 & 1 & 2 & 0 \\ 1 & 0 & 0 & 1 \\ 0 & 1 & 0 & 0 \end{bmatrix}$$

$$= \begin{bmatrix} 1 & 2 & 1 & 1 \\ 1 & 1 & 2 & 1 \\ 1 & 1 & 0 & 1 \\ 0 & 1 & 1 & 0 \end{bmatrix},$$

$$(A + A^2) + A^3 = \begin{bmatrix} 1 & 2 & 1 & 1 \\ 1 & 1 & 2 & 1 \\ 1 & 1 & 0 & 1 \\ 0 & 1 & 1 & 0 \end{bmatrix} + \begin{bmatrix} 1 & 1 & 2 & 1 \\ 1 & 2 & 0 & 1 \\ 0 & 1 & 2 & 0 \\ 1 & 0 & 0 & 1 \end{bmatrix}$$

$$= \begin{bmatrix} 2 & 3 & 3 & 2 \\ 2 & 3 & 2 & 2 \\ 1 & 2 & 2 & 1 \\ 1 & 1 & 1 & 1 \end{bmatrix},$$

The resulting sum matrix has no 0-entries. Thus, every vertex can be reached from every vertex. Such a graph is called **strongly connected.**

The same process works for an unconnected graph, such as H of Figure 6.27.

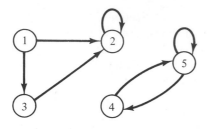

	1	2	3	4	5
1	0	1	1	0	0
2	0	1	0	0	0
B = 3	0	1	0	0	0
4	0	0	0	0	1
5	0	0	0	1	1

(*i*) A Directed Graph H   (*ii*) The Adjacency Matrix B of H

FIGURE 6.27

The reader may wish to verify the following calculations.

$$B^2 = \begin{bmatrix} 0 & 2 & 0 & 0 & 0 \\ 0 & 1 & 0 & 0 & 0 \\ 0 & 1 & 0 & 0 & 0 \\ 0 & 0 & 0 & 1 & 1 \\ 0 & 0 & 0 & 1 & 2 \end{bmatrix}, \quad B^3 = \begin{bmatrix} 0 & 2 & 0 & 0 & 0 \\ 0 & 1 & 0 & 0 & 0 \\ 0 & 1 & 0 & 0 & 0 \\ 0 & 0 & 0 & 1 & 2 \\ 0 & 0 & 0 & 2 & 3 \end{bmatrix},$$

$$B^4 = \begin{bmatrix} 0 & 2 & 0 & 0 & 0 \\ 0 & 1 & 0 & 0 & 0 \\ 0 & 1 & 0 & 0 & 0 \\ 0 & 0 & 0 & 2 & 3 \\ 0 & 0 & 0 & 3 & 5 \end{bmatrix}.$$

$$B + B^2 + B^3 + B^4 = \begin{bmatrix} 0 & 7 & 1 & 0 & 0 \\ 0 & 4 & 0 & 0 & 0 \\ 0 & 4 & 0 & 0 & 0 \\ 0 & 0 & 0 & 4 & 7 \\ 0 & 0 & 0 & 7 & 11 \end{bmatrix}.$$

The results indicate: (a) Even though vertices 2 and 3 can be reached from 1, 1 cannot be reached from either 2 or 3. (b) The graph H is not strongly connected. (c) The subgraph with vertices 4 and 5 and all the edges incident to them is isolated; so is its complement (the subgraph with vertices 1, 2 and 3 and all the edges incident to them), and hence H is not connected.

6.15 ROOTED TREES

[The definitions and basic results on trees from Section 6.7 are assumed here.]

A tree is a connected graph with no cycles. Its edges have no orientation. However, once any one vertex is designated as a root, a first or original vertex, every edge acquires a direction as a result, as is illustrated in Figure 6.28.

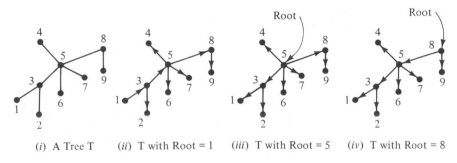

(*i*) A Tree T   (*ii*) T with Root = 1   (*iii*) T with Root = 5   (*iv*) T with Root = 8

FIGURE 6.28

6.28 Definition:[1]   A **rooted tree** is a graph without cycles, with a designated vertex, called a **root**. A rooted tree is denoted by a triple $T = (V,E,r)$, where V and E denote the usual vertex set and edge set, and where $r \in V$ is the root. Such a rooted tree is also called a **tree with root** r. (This notation is convenient but not universal.)

A rooted tree is thus a directed graph, with each edge directed "away" from the root. Since, by Theorem 6.18 (Section 6.7), there is a unique path between the root and any given vertex v, the last edge in that path from the root to v has v itself

---

[1]For a recursive definition of a rooted tree, see Section 17.3.

as the terminal vertex; however, in all other edges incident with
v, *v* in the *initial* vertex.

> ***6.29 Definition:***   Let $T = (V,E,r)$ be a rooted tree, and
> let $u \in V$. Every vertex *v* of *T*, such that (u,v) is a directed
> edge of *T*, is called a **son** of *u*. For every vertex *w* of *T*, *w* is a
> **descendant** of *v* iff there is a directed path from *v* to *w*.
>
> A subgraph $S = (U,F,w)$ is a **subtree** of the rooted tree
> *T* iff *S* is a rooted tree with root *w*, the vertices in *U* are *w*
> and all its descendants in *T*, and the edges in *F* are all the
> edges of *T* both of whose vertices are in *U*.
>
> A null subtree of a single vertex, i.e. a subtree with no
> other (proper) subtrees, is called a **leaf** of the tree.

The example of Figure 6.29 shows a tree with root 5. The
sons of 5 are 3, 4, 6, 7 and 8. In addition, 5 has descendants 1, 2
and 9. The subtrees in part (ii) of the figure include the leaves
4, 6 and 7. They are null trees (with no edges) and with only
one vertex. The remaining leaves of *T* are 1, 2 and 9. (No
confusion arises from considering a leaf both as a vertex and as
a subtree.)

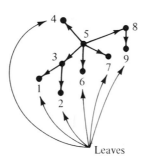

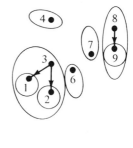

(*i*) The Tree T with Root 5   (*ii*) Subtrees of T

FIGURE 6.29

The reader should not overlook the fact that not all
directed trees are rooted trees. A **directed tree** is any tree, each
of whose edges is assigned a direction. Such an assignment
need not produce a root; instead, it might cause one vertex to be
the terminal vertex in more than one directed edge—an
arrangement impossible in a rooted tree.

6.30 Theorem: In a rooted tree $T = (V,E,r)$, there is no edge in which the root r is the terminal vertex. Every vertex other than the root is terminal in exactly one edge.

The reader should be cautioned that trees are often depicted with undirected edges; the directions are implied by placing the root of the tree at the top of the diagram, and the sons of each vertex are placed lower in the diagram than the vertex itself. The tree T of Figure 6.30(i) then appears as is shown in Figure 6.30(ii).

In such a representation, the leaves of the tree always appear at the bottom of the diagram, but at differing heights.

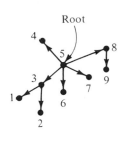

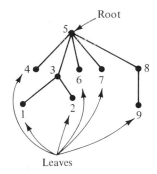

(*i*) The Tree T with Root 5   (*ii*) A Common Representation of T

FIGURE 6.30

6.16 BINARY TREES

An **ordered tree** is a rooted tree in which, for every vertex with sons, the sons are placed in order so that there is a first son, a second, etc.

Although a **binary tree** may be taken as just a tree in which no vertex has more than two sons, most of the computer scientific applications of binary trees make use of the fact that there are at most a **left son** and a **right son** for any vertex.

The tree in Figure 6.31 is a binary tree, even though vertex 9 has only one son. In such a case, the son may be designated as either the left son or the right one. The subtree whose root is 2

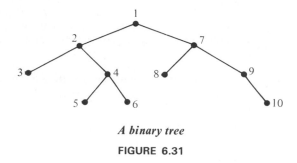

A binary tree

FIGURE 6.31

is a **complete binary tree,** since every vertex which is not a leaf has two sons.

A **binary tree** then is ordered, where customarily the left son precedes the right one in the order.

In many computer applications of binary trees, it is required to visit all vertices of the binary tree systematically. There are three common orders for traversing a binary tree:

Pre-order traversal: In the example of Figure 6.32, the pre-order traversal dictates the following visiting order of the vertices, 1, 2, 4, 5, 7, 9, 11, 12, 10, 3, 6, 8.

The rule for pre-order traversal is: *For any subtree, first visit the root, next the entire left subtree, and last the right subtree.* This may be regarded as a "top-left-right" order.

In-order and end-order traversals are illustrated in Figure

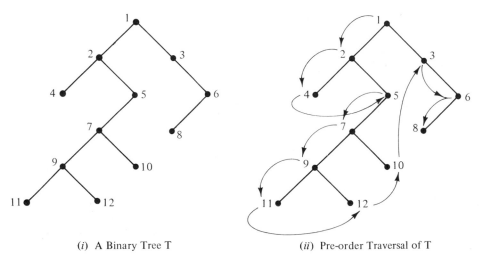

(*i*) A Binary Tree T   (*ii*) Pre-order Traversal of T

FIGURE 6.32

6.33(i) and (ii), respectively.

The **in-order**[1] **traversal** of T encounters the vertices in the order, 4, 2, 11, 9, 12, 7, 10, 5, 1, 3, 8, 6. The rule for in-order traversal is: *For any subtree, first visit its entire left subtree, next visit the root, and last visit the right subtree.* This may be regarded as a "left-top-right" order.

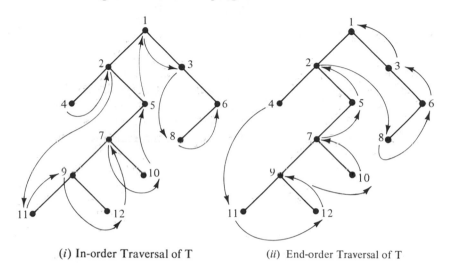

(*i*) In-order Traversal of T   (*ii*) End-order Traversal of T

FIGURE 6.33

The **end-order**[2] **traversal** of T encounters the vertices in the order, 4, 11, 12, 9, 10, 7, 5, 2, 8, 6, 3. The rule for end-order traversal is: *For any subtree, first visit its entire left subtree, next visit its entire right subtree, and last visit the root.* This may be regarded as a "left-right-top" order.

All three traversal orders are defined recursively (see Chapter 17) and are therefore economical to program. There are, of course, other traversal orders, but these three are the most commonly used.

6.17 THE TREE OF AN EXPRESSION

The expression $(((a + b) * c) * (d - e)) - (f - (g * h))$ may be represented as the (ordered) binary tree

---

[1]In-order is sometimes called *post-order* and *symmetric-order*.
[2]End-order is sometimes called *post-order* and *bottom-up order*.

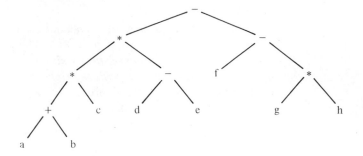

Before it can be evaluated, each letter name must be assigned a value. Also, before each subtree can be evaluated, both its left and right subtrees must have been evaluated.

For example, to evaluate $(((3 + 5) * 2) * (7 - 4)) - (12 - (2 * 3))$ from its binary tree representation

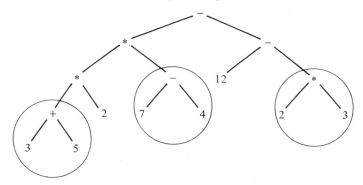

only the circled subtrees are ready to be evaluated. We arbitrarily choose the leftmost

and replace it with the result 8 where the root of the subtree used to be. The result then is

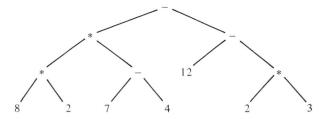

We may say that the tree has been **collapsed, resolved** or **pruned.**

We again may evaluate any of three subtrees and again we choose the leftmost

and replace it with 16. The result of the latest collapse is

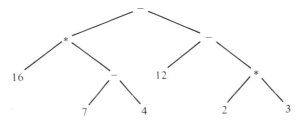

in which the leftmost of all subtrees which are ready for evaluation is

If we continue to evaluate the leftmost of all the ready subtrees, the remaining steps of the evaluation are,

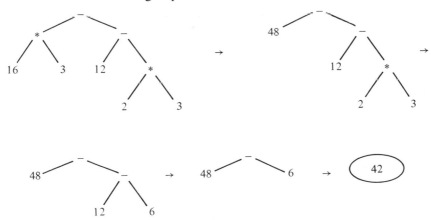

A similar treatment may be given to the "logical expression"

$$(((P \Rightarrow Q) \ \& \ (Q \Rightarrow R)) \Rightarrow (P \Rightarrow R)).$$

After assigning particular truth values (see Section 16.7) to the variables, $P \leftarrow T$, $Q \leftarrow F$ and $R \leftarrow F$, the binary tree to be evaluated is

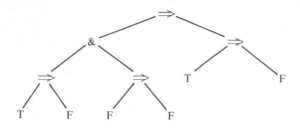

The evaluation procedure may be abbreviated by **climbing the tree,** rather than collapsing it and copying the resulting tree each time. In climbing the tree, the evaluation takes place in the same fashion described above, but the result is simply entered by the root of the evaluated subtree. The direction of climbing is from bottom up, toward the root; when the root is assigned a value, the entire binary tree has been evaluated. (The same technique, of course, may be used with the preceding example.)

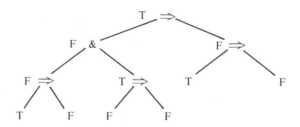

6.18 DERIVATION TREES

A **grammar,** a basic construct in formal language theory, may be loosely regarded as a system of **rewriting rules** whose intention is to derive sentences in the following way: Every **derivation** begins with the starting symbol S (sometimes called the "sentence symbol"). Any one of the rules whose left-hand side is S may next be used. For example, suppose that

$$S \rightarrow ART\ NP\ VP$$

is one of the rewriting rules we may use in our grammar. (*ART*, *NP* and *VP* may be interpreted here as "article," "noun-phrase," and "verb-phrase," respectively.) S is then allowed to be replaced by the ordered triple *ART NP VP*. The derivation begins with the replacement of the first row by the second:

$$S$$
$$ART\quad NP\quad VP.$$

In the following rewriting rules, we include more than we intend to use to illustrate the derivation of one particular sentence. The additional rules, however, may aid the understanding and provide experimentation room.

1. $S \rightarrow ART\ NP\ VP$	9. $V \rightarrow$ walked
2. $S \rightarrow NP\ VP\ NP$	10. $V \rightarrow$ is drinking
3. $NP \rightarrow N$	11. $N \rightarrow$ lion
4. $NP \rightarrow ART\ NP$	12. $N \rightarrow$ chair
5. $NP \rightarrow PREP\ NP$	13. $ART \rightarrow$ the
6. $NP \rightarrow ADJ\ N$	14. $ADJ \rightarrow$ fierce
7. $VP \rightarrow V$	15. $ADV \rightarrow$ majestically
8. $VP \rightarrow V\ ADV$	16. $PREP \rightarrow$ on

The understanding is that any of the rules may be used any time its left-hand side (to the left of the arrow) appears in the current derived string. Thus, starting with S, we may obtain the following sequence of strings, leading to the sentence in the final line. The single rule by which each line is obtained from its predecessor appears on the left of the line.

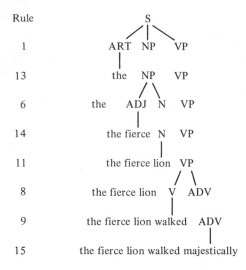

Rule	
1	S → ART NP VP
13	the NP VP
6	the ADJ N VP
14	the fierce N VP
11	the fierce lion VP
8	the fierce lion V ADV
9	the fierce lion walked ADV
15	the fierce lion walked majestically

The individual rewriting rules used in the derivation are illustrated by the ordered tree of Figure 6.34. It is called a **derivation tree** for the sentence "the fierce lion walked majestically." The sentence may be read from the tree as the succession of leaves from left to right, disregarding their heights.

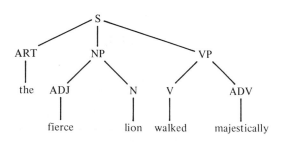

A derivation tree

FIGURE 6.34

The reader may wish to construct a derivation tree for the sentence "the wet chair is drinking slowly on the lion."

As another example, consider the rewriting rules

1. $S \rightarrow V$
2. $V \rightarrow (V B V)$
3. $V \rightarrow (\sim V)$
4. $V \rightarrow P$
5. $V \rightarrow Q$
6. $V \rightarrow R$
7. $B \rightarrow \Rightarrow$
8. $B \rightarrow \&.$

Figure 6.35 shows a derivation tree for the propositional form

$$(((P \Rightarrow Q) \,\&\, (Q \Rightarrow R)) \Rightarrow (P \Rightarrow R)).$$

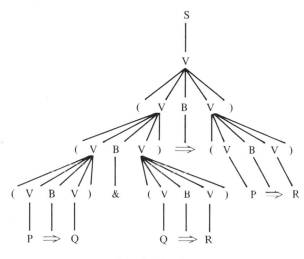

A derivation tree

FIGURE 6.35

The reader may attempt to obtain a derivation tree for any propositional form which uses only the variables P, Q and R, in its equivalent form restricted to the logical operations in the rewriting rules.

6.19 DIRECTED LABELED GRAPHS; FINITE AUTOMATA

When labels are assigned to the edges of a directed graph, the result in a **directed labeled graph.** There exist various reasons for labeling the edges, and each reason determines the manner of labeling.

The example in Figure 6.36 is a **state diagram** of a **sequential machine,** a variety of a **finite automaton.** The labels on the edges are of the form x/y, where x denotes the **input** digit and y the **output** digit to the sequential machine. The vertices of the sequential machine are called **states.**

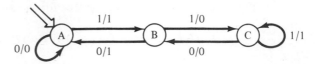

State diagram of a sequential machine: A binary 3-multiplier

FIGURE 6.36

The particular sequential machine illustrated in Figure 6.36 is a binary 3-multiplier. A string of binary digits, which is input to the multiplier, is regarded as a binary integer, read from right to left; that is, starting with the least significant digit. (Binary arithmetic is discussed in Chapter 14.) The first input digit is always applied to the **initial state** A. (The initial state is marked with a double arrow.) Each input digit in turn causes a one-step move along the arrow labeled by it. It also causes the deposit of an output digit, which is to the right of the "/" in the label.

We trace the input string 001101, which is the binary representation of 13 with leading zeros attached (to allow the rest of the outputs to exit the multiplier after running out of input digits). The first input 1 is applied to the initial state A. The edge labeled 1/1 from A to B is the one followed, because the input digit is 1. Thus, the **next state** is B and the output is 1. The next digit of the input string is 0. When it is applied to the state B, the edge labeled 0/1 from B to A is the one followed, making A the next state and 1 the output. The entire succession of next states and outputs is shown in Table 6.1.

The resulting output sequence is 100111, which is the binary representation of $39 = (3)(13)$.

TABLE 6.1

Applying 001101 to the Binary
3-Multiplier.

State	Input	Next State	Output
A	1	B	1
B	0	A	1
A	1	B	1
B	1	C	0
C	0	B	0
B	0	A	1

The reader may wish to trace other input sequences and to verify that the output integer is triple that of the input integer.

Another, related application of directed labeled graphs is shown in Figure 6.37. It is an example of a **nondeterministic finite-state acceptor,** another variety of a finite automaton. A string of binary digits, called the **input string,** is traced (as in the sequential machine of the preceding example) starting with the initial state A. (In an acceptor, it is customary to read the input string from left to right.) An edge is followed if its label matches the present input.

The state B is enclosed with a double circle, which indicates that it is an **accepting state;** that is, it **accepts** an input string which leads to it. The states A and C are not accepting states. An input string x is **accepted** iff there exists a path leading from the initial state to an accepting state so that the sequence of labels of the traversed edges, in order, is precisely the input string x.

For example, the sequence 100 (read from left to right) is accepted along the path $A \xrightarrow{1} B \xrightarrow{0} C \xrightarrow{0} B$, even though another path, $A \xrightarrow{1} B \xrightarrow{0} A \xrightarrow{0} C$, is properly labeled but does not lead to B. On the other hand, there is no path which leads from A to B and is labeld 1010; thus, this input string is not accepted.

Such a device is used to **recognize** sets of strings (to tell which strings belong to the set) which are **regular;** that is, sets which can be so recognized by an acceptor of this kind with a *finite* number of states.

Many other applications of directed labeled graphs are common in computer science. The constructs illustrated by such graphs are varied and subjects of entire fields of study.

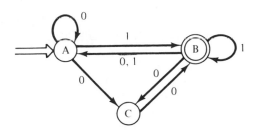

A nondeterministic finite-state acceptor for strings ending either with 1 or with 00

FIGURE 6.37

7

Relations

7.1 WHAT ARE RELATIONS LIKE? EXAMPLES

The concepts of ordering, function, partition, and Cartesian product, among others, may be treated as relations. Relations may be defined as sets which have special structure. Before presenting a formal treatment of relations, an example is presented to aid the reader's intuition.

The relation (in the intuitive sense) is between students of a particular university and the courses the students are enrolled in. We then consider certain pairs of objects, each pair consists of a student and a course the student takes. The *relation* is the *set* of all such *ordered pairs*. The fact should not be overlooked that there are (probably) several ordered pairs with the same left member, since the same student is likely to take more than one class; for example, the enrollment record for J. Cox for a certain term may consist of Eng. 101, Math. 310, Hist. 222, and P.E. 120 and thus the ordered pairs (J. Cox, Eng. 101), (J. Cox, Math. 310), (J. Cox, Hist. 211), and (J. Cox, P.E. 120) *are elements of the relation*. It should also be realized that there are likely to be more than one student in a course and thus there will be several ordered pairs with the same right member. For example, each name in the class roster for Eng. 101 serves as a left member for an ordered pair in the relation, with Eng. 101 as its right member.

It is important to realize that the relation just described is not the "connection" between the students and their classes, nor the "relationship" between the two, but the collection (set) of all ordered pairs (or pairings), each of which has as a first member, or first coordinate, a student's name and as a second coordinate the name of a course the student is taking.

If we were to envision all students of the university listed down the left of a gigantic sheet of paper, and all the courses offered this term listed along the top of that sheet, we could place a mark in the row of a student and a column of a course taken by the student to indicate the ordered pair. If this were done for all students and all courses taken by them, the resulting chart (**table**) would represent the relation exemplified above. (See Figure 7.1, below, for a smaller example.)

Another example of a relation, close to the preceding one, is the set of all ordered pairs consisting of a student and the first course the student takes in the week. Here, for each student there is exactly one ordered pair in the relation.

A third example is the relation "between groceries and their prices," that is, the set of all ordered pairs whose first coordinate is an item of grocery and whose second one is the price of that item.

7.2 DOMAIN, RANGE AND CO-DOMAIN

An example of a relation need not be a familiar one, nor must it come from the familiar surroundings. All we need in order to present an example of a relation is a collection of left members (student names, grocery items, in our examples), called the **domain** of the relation, a collection of right members (course names, grocery prices), called the **range,** or **co-domain**[1] of the relation, and an indication of the pairing of members of the domain with those of the range. No further "meaning" need be attached to these "pieces of the game."

7.3 THE TABLE OF A RELATION

To illustrate, we may simply specify some letters as the members of the domain, say a, b, c, and d, and some numbers as the members of the range, say 1, 2, 3, 4, 5, and indicate by a

---

[1]Range and co-domain are not identical in their usage, although sometimes they are used interchangeably.

table which letters **are related to** which numbers, as is shown in Figure 7.1.

R	1	2	3	4	5
a	x		x		
b		x	x		x
c	x			x	
d		x			

$$\begin{bmatrix} 1 & 0 & 1 & 0 & 0 \\ 0 & 1 & 1 & 0 & 1 \\ 1 & 0 & 0 & 1 & 0 \\ 0 & 1 & 0 & 0 & 0 \end{bmatrix}$$

(*i*) Table Representation of a Relation R   (*ii*) Boolean Matrix of R

FIGURE 7.1

The relation, which is called R here, is in fact the set of eight ordered pairs indicated by the x-marks in Figure 7.1(i). That is,

$$R = \{(a,1), (a,3), (b,2), (b,3), (b,5), (c,1), (c,4), (d,2)\}.$$

We also use the language "*a* is *R*-**related to** 1" and write "*a R* 1" to record this fact. The **Boolean matrix** of the relation R, or simply the **matrix** of R, is very similar to the table of R. The only difference between the two, as illustrated in Figure 7.1(ii), is that 1's appear in place of the x's and 0's appear in place of the empty entries. The Boolean matrix of a relation has interesting uses, as seen below, but is more cumbersome to handle than is the table of the relation.

Since no particular meaning was attached to the "connections" between the letters and the numbers, we could have used *any* scheme of x-marks in the matrix of Figure 7.1 and the resulting collection of ordered pairs would be a relation. In fact, we do so in Figure 7.2 for the relation we call T, where D is the domain and C the co-domain.

T	1	2	3	4	5
a		x			
b					
c					x
d					

$T = \{(a, 2), (c, 5)\}$
$D = \{a, b, c, d\}$
$C = \{1, 2, 3, 4, 5\}$

Representation of a relation T

FIGURE 7.2

7.4 THE EXTREMES: THE CARTESIAN PRODUCT AND THE EMPTY RELATION

No matter what two nonempty sets are used for domain and co-domain, two relations are always present. The first is the **universal relation** U. It is the collection of all ordered pairs possible with first coordinate from the domain and second coordinate from the range. The table of such a relation is full of x-marks, as is shown in Figure 7.3(i), since every letter is U-related to every number. Similarly, the Boolean matrix of U is full of 1's. The universal relation is commonly called the **Cartesian product** of D and C.

U	1	2	3	4	5
a	x	x	x	x	x
b	x	x	x	x	x
c	x	x	x	x	x
d	x	x	x	x	x

$$\begin{bmatrix} 1 & 1 & 1 & 1 & 1 \\ 1 & 1 & 1 & 1 & 1 \\ 1 & 1 & 1 & 1 & 1 \\ 1 & 1 & 1 & 1 & 1 \end{bmatrix}$$

(i) Table   (ii) Boolean Matrix

The universal relation—the Cartesian product

FIGURE 7.3

7.1 Definition: Let A and B be sets. Then the **Cartesian product** of A and B is $A \times B = \{(a,b): a \in A \text{ and } b \in B\}$.

It then follows that every relation with D and C as domain and co-domain, respectively (intuitively, any scheme of x-marks in the table), is a subset of the universal relation, or the Cartesian product of D and C.

The other relation which always exists for any nonempty domain and co-domain is the **null** or **empty relation.** It is simply the empty set of ordered pairs and its table has no x-mark. In other words, no member of the domain is related to any member of the co-domain. (The Boolean matrix is all zeros.)

In between the two extremes, the Cartesian product and the empty relation, there are as many different relations as there are different schemes of placing x-marks in the matrix. Some such relations have special properties, which are encountered below.

7.5 THE FORMAL DEFINITION

Before proceeding with the discussion, the concepts encountered thus far are formalized.

> **7.2 Definition:**   Let D and C be nonempty sets. A **relation** R on D to C is a subset of $D \times C$.
>   We say that $x \in D$ is R-**related** to $y \in C$, and we write $x \, R \, y$, iff $(x,y) \in R$.
>   The **domain** of R is
> domain $R = \{x \in D$: for some $y \in C, (x,y) \in R\}$.
>   The **range** of R is
> range $R = \{y \in C$: for some $x \in D, (x,y) \in R\}$.

7.6 THE INVERSE OF A RELATION

There are two further concepts concerning relations, which are encountered less frequently in the literature, but which may enhance the reader's view of the subject: those of the inverse of a relation and the composition of relations.

R^{-1}

The **inverse** of a relation R, denoted by R^{-1}, is obtained by reversing each of the pairs of R. As an example, the inverse of R of Figure 7.1 has the representation shown in Figure 7.4, with the domain of R serving as the range of R^{-1} and the range of R as the domain of R^{-1}.

R	1	2	3	4	5
a	x		x		
b		x	x		x
c	x			x	
d		x			

R^{-1}	a	b	c	d
1	x		x	
2		x		x
3	x	x		
4				x
5		x		

(*i*) Table for R   (*ii*) Table for R^{-1}

A relation R and its inverse R^{-1}

FIGURE 7.4

7.3 Definition: The **inverse** of a relation R is

$$R^{-1} = \{(x,y): (y,x) \in R\}.$$

Thus, $x \, R \, y$ iff $y \, R^{-1} \, x$. Also, $(A \times B)^{-1} = B \times A$, for all sets A and B.

7.7 COMPOSITION OF RELATIONS

Two relations are illustrated in Figure 7.5, relation R on A to B and relation S on B to C. Since $(c,5) \in R$ and $(5,x) \in S$, the member 5 of B may serve to make $c \in A$ related to $x \in C$ by a new relation we call the **composition** of R and S.

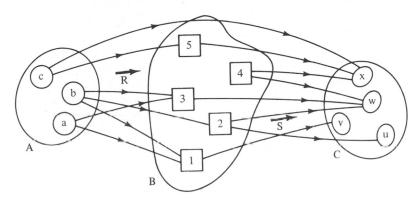

Composition of two relations, R on A to B and S on B to C

FIGURE 7.5

For the time being, we call the new relation T. The three relation matrices are shown in Figure 7.6. We say that b is T-related to w, and write $b\,T\,w$, iff there exists an element n of B such that $b\,R\,n$ and $n\,S\,w$; indeed, $n-2$ is such an element of B.

R	1	2	3	4	5
a	1	0	1	0	0
b	1	1	1	0	0
c	0	0	0	0	1

S	u	v	w	x
1	0	1	0	0
2	1	0	1	0
3	0	0	1	0
4	0	0	1	1
5	0	0	0	1

T	u	v	w	x
a	0	1	1	0
b	1	1	1	0
c	0	0	0	1

[R]   [S]   [T]

The Boolean matrices of R, S and T

FIGURE 7.6

Before we define composition of relations and present a notation for it, the reader must be made aware of a danger.

[*WARNING TO THE READER: There exists a discrepancy between the common notation for composition of relations and the common notation for composition of functions. (See Section 9.4.) The order of the two symbols for the relations* R *and* S, *which describes the composition of these relations, is often given as* RS *or* R · S. *(A probable reason for this may be seen in the computation of the relation matrix of the composition, Figure 7.7.) If, however,* R *and* S *were functions, that same composition would be written* S ∘ R, *read "*S *of* R*," and the image on a member* x *of the domain of* S *by the composition would be written as* (S ∘ R) (x) = S(R(x)). *This notation for the composition of functions is used often enough for the composition of relations to cause confusion by the unaware. In Definition 7.4, below, we use both notations, but unless indicated by the multiplication of matrices, we prefer the notation to conform with that for the composition of functions. In that order, we shall always use the little circle,* ∘, *to denote the composition of two relations.*]

7.4 Definition: Let *A*, *B* and *C* be nonempty sets and let *R* be a relation on *A* to *B* and *S* a relation on *B* to *C*.

The **composition** of *R* and *S*, denoted by *RS* or *S* ∘ *R*, is a relation on *A* to *C* defined by: $\forall x \in A$ and $\forall z \in C$, *x* (*RS*) *z* (also *x* (*S* ∘ *R*) *z*) iff there is $y \in B$ such that *x R y* and *y S z*; i.e., $(x,z) \in RS$ (also $(x,z) \in S \circ R$) iff there is $y \in B$ such that $(x,y) \in R$ and $(y,z) \in S$.

The Boolean matrix of the composition of *R* and *S* may be computed as the **Boolean product** of the two matrices for *R* and *S*. (Matrix multiplication is illustrated in Section 6.14.) **Boolean matrix multiplication** differs from the common matrix multiplication only in the addition part of the operation: the Boolean sum of 1 and 1 is 1,

$$1 + 1 = 1.$$

We illustrate with the computation of the entry in row *b* and column *w* of the matrix of the composition, shown in Figure 7.6 as [*T*]. This entry is the (Boolean matrix) product of row *b* of the matrix [*R*] of the relation *R* and column *w* of the matrix [*S*] of the relation *S*. (Both matrices are shown in Figure 7.6.) The product is computed as follows:

$$[1 \quad 1 \quad 1 \quad 0 \quad 0] \cdot \begin{bmatrix} 0 \\ 1 \\ 1 \\ 1 \\ 0 \end{bmatrix} = \begin{aligned} &(1)(0) + (1)(1) + (1)(1) \\ &\quad + (0)(1) + (0)(0) \\ &= 0 + 1 + 1 + 0 + 0 = 1. \end{aligned}$$

We continue to denote the matrix of any relation R by $[R]$. Then, the product of the two matrices in our example, the product of $[R]$ and $[S]$, is shown in Figure 7.7 as $[RS]$.

$$\begin{bmatrix} 1 & 0 & 1 & 0 & 0 \\ 1 & 1 & 1 & 0 & 0 \\ 0 & 0 & 0 & 0 & 1 \end{bmatrix} \cdot \begin{bmatrix} 0 & 1 & 0 & 0 \\ 1 & 0 & 1 & 0 \\ 0 & 0 & 1 & 0 \\ 0 & 0 & 1 & 1 \\ 0 & 0 & 0 & 1 \end{bmatrix} = \begin{bmatrix} 0 & 1 & 1 & 0 \\ 1 & 1 & 1 & 0 \\ 0 & 0 & 0 & 1 \end{bmatrix}$$

$$[R] \cdot [S] = [RS]$$

FIGURE 7.7

The order RS in the notation for the composition of the relations corresponds to the order of the matrices in the customary manner of matrix multiplication.

The laws of operating with relations are numerous and they form an extensive calculus. Here, we present only three results which are helpful in understanding and manipulating relations. Of course, they hold true when the relations are functions, but the results are presented here in the more general setting. The first states that the inverse of the inverse of a relation is the relation itself and that the inverse of a composition of relations is the composition of the inverses in the *reverse order*.

7.5 Theorem: Let R and S be relations. Then

(i) $(R^{-1})^{-1} = R$;

(ii) $(S \circ R)^{-1} = R^{-1} \circ S^{-1}$.

A point of caution: $R^{-1} \circ R$ is *not* 1, unity, identity or anything similar. The matrices of R and R^{-1} from Figure 7.4

are sufficient to make the point. The product $[R^{-1} \circ R] = [RR^{-1}] = [R] \cdot [R^{-1}]$ is,

$$\begin{bmatrix} 1 & 0 & 1 & 0 & 0 \\ 0 & 1 & 1 & 0 & 1 \\ 1 & 0 & 0 & 1 & 0 \\ 0 & 1 & 0 & 0 & 0 \end{bmatrix} \cdot \begin{bmatrix} 1 & 0 & 1 & 0 \\ 0 & 1 & 0 & 1 \\ 1 & 1 & 0 & 0 \\ 0 & 0 & 1 & 0 \\ 0 & 1 & 0 & 0 \end{bmatrix} = \begin{bmatrix} 1 & 1 & 1 & 0 \\ 1 & 1 & 0 & 1 \\ 1 & 0 & 1 & 0 \\ 0 & 1 & 0 & 1 \end{bmatrix}.$$

The main-diagonal entries of the product are all 1, but there the similarity to an identity, or a unit, ends.

It may be of interest to note that, for any nonempty relation R on A to B, the composition $RR^{-1} = R^{-1} \circ R$ is a relation on A to A, and the composition $R^{-1}R = R \circ R^{-1}$ is a relation on B to B. Both are *equivalence relations* (Section 7.16), a fact which accounts for several features of the product matrix, including the 1's on the main diagonal and the symmetry of the 1-entries about the main diagonal.

For the next theorem we use the following notation. Where A is a subset of the domain of a relation R, $R(A)$ denotes the set of all R-**relatives** of members of A; that is, the set of all members of the range of R which are R-related to some members of A. Formally,

$$R(A) = \{y: x \, R \, y \text{ for some } x \in A\}.$$

$R(A)$ may be intuitively thought of as the set of R-images of members of A.

The next theorem states that *composition of relations is associative*. It also states that the set of images of a set by the composition of relations may be obtained one step at a time.

7.6 Theorem: Let R, S, and T be relations and let A be a set. Then,

(i) $R \circ (S \circ T) = (R \circ S) \circ T$;
(ii) $(R \circ S)(A) = R(S(A))$.

The next theorem states that the image set of a union (alt., intersection) is the union (alt., subset of the intersection) of the image-sets.

7.7 Theorem: Let R be a relation and let A and B be sets. Then,

(i) $R(A \cup B) = R(A) \cup R(B)$;

(ii) $R(A \cap B) \subseteq R(A) \cap R(B)$.

This theorem is easily generalized to any number of sets, or a collection of sets indexed by an arbitrary indexing set.

It should be noted in passing that relations in general, and Cartesian products in particular, apply to dimensions higher than two; thus, an *n*-**ary relation** is a subset of the *n*-**ary Cartesian product** $A_1 \times A_2 \times \ldots \times A_n$ and is a set of ordered *n*-tuples of the form $(a_1, a_2, \ldots a_n)$, with each coordinate a member of the corresponding set. In that sense, we have been considering only **binary relations** and **binary Cartesian products** and shall continue to do so, to satisfy the foreseeable needs of the computer scientist. The extension to higher dimensions, when and if it becomes necessary, is fairly obvious.

7.8 RELATIONS ON A SET

Relations are of special interest when their domain and co-domain are the same set. When the relation R is from the set A to A, i.e., $R \subseteq A \times A$, we say that R is a **relation on** A. For the remainder of this chapter, we consider some special relations of this sort which occur frequently.

(In truth, we are not abandoning the general case, for we could suppose that R is a relation and that A is the set of all members of the domain and range of R, i.e. $A = $ (domain R) $\cup$ (range R). We may then regard R as a relation on the set A.)

7.9 THE GRAPH OF A RELATION

[*Graphs are discussed in Chapter 6. The concepts and basic definitions are assumed here, but the terminology is largely self-explanatory, so that the reader need not refer to the coverage in Chapter 6 unless particular questions arise. If reference is desired, Section 6.11, in particular, illustrates the graph of a relation on a set.*]

Let $A = \{a,b,c\}$ and let the relation R on A be defined by the table or the Boolean matrix of Figure 7.8. The directed graph of part (iii) of the figure is called the **graph of the relation R**. It has the members of the set A as **vertices**. Also, it has a **directed edge** from vertex v to vertex u iff $(v,u) \in R$, for any v and u in A.

R	a	b	c
a		x	
b	x		x
c	x		x

(i) Table of R

$$\begin{array}{c|ccc} & a & b & c \\ \hline a & 0 & 1 & 0 \\ b & 1 & 0 & 1 \\ c & 1 & 0 & 1 \end{array}$$

(ii) Boolean Matrix of R

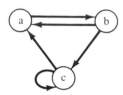

(iii) Directed Graph of R

FIGURE 7.8

7.8 Definition: Let A be a nonempty set and let R be a relation on A. The **graph of R** is the directed graph $G_R = (V,E)$, where $V = A$ and $E = R$. I.e., the set of vertices V of the graph is the set A, and the directed edges of the graph are the (ordered) pairs which are members of the relation R.

7.10 THE RELATION OF A DIRECTED GRAPH

Given the directed graph in part (iii) of Figure 7.8, above, the **relation of the directed graph** is defined in parts (i) and (ii). In general, the set A on which the relation is defined is the set of vertices of the graph; moreover, an ordered pair (x,y) is in the relation (x is related to y) iff there is a directed edge in the graph from vertex x to vertex y.

7.9 Definition: Let $G = (V,E)$ be a directed graph. The **relation of G** is the relation R_G defined on the vertex set V by:

$$\forall x,y \in V, (x,y) \in R_G \text{ iff } (x,y) \in E.$$

7.11 RELATION, GRAPH AND THE ADJACENCY MATRIX

From the preceding two sections, it follows that every relation R on a nonempty set A determines uniquely a directed graph G_R, just as every directed graph $G = (V,E)$ determines uniquely a relation R_G on V. (Only the *diagram* of G_R may have more than one appearance, but the abstract structure of the graph is indeed uniquely determined.)

The **adjacency matrix** of the graph (see Section 6.13) is precisely the Boolean matrix of the corresponding relation.

7.12 IDENTITY AND REFLEXIVE RELATIONS

In addition to the universal relation and the empty relation, described in Section 7.4, a frequently used relation on a set is the **identity relation,** also called the **diagonal relation** for reasons obvious from Figure 7.9. It is defined as

$$Id(A) = \{(x,x): x \in A\}.$$

The only element related to x is x itself.

Id	a	b	c	d
a	x			
b		x		
c			x	
d				x

$$\begin{bmatrix} 1 & 0 & 0 & 0 \\ 0 & 1 & 0 & 0 \\ 0 & 0 & 1 & 0 \\ 0 & 0 & 0 & 1 \end{bmatrix}$$

(*i*) Table of the Identity Relation   (*ii*) The Graph G of the Identity Relation   (*iii*) The Adjacency Matrix of G

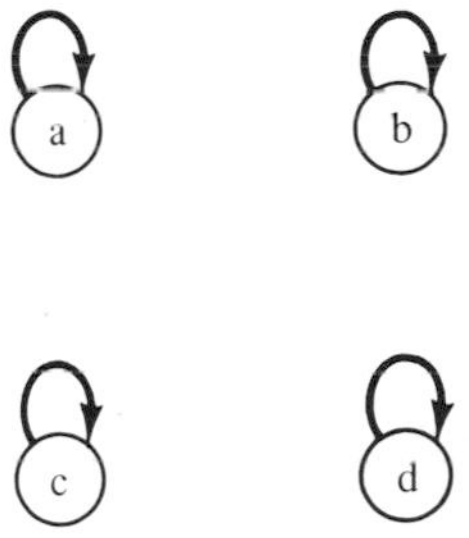

FIGURE 7.9

The graph of the identity relation is a **pseudograph** (an edge connecting a vertex to itself is allowed); its edges are exactly those connecting each vertex to itself.

7.10 Definition: Let A be a set and let R be a relation on
A. Then R is **reflexive** iff $(x,x) \in R$ for every $x \in A$.
A relation which is not reflexive is called **irreflexive**.[1]

Thus, the identity relation (Figure 7.9) is reflexive, as are
the relations illustrated in Figure 7.10.

R_1	a	b	c
a	x	x	
b		x	
c			x

R_2	a	b	c
a	x	x	x
b		x	x
c			x

R_3	a	b	c
a	x		x
b	x	x	x
c		x	x

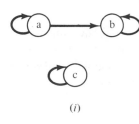

(i)

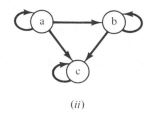

(ii)

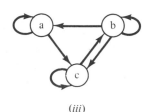

(iii)

Reflexive relations and their graphs

FIGURE 7.10

The graph of every reflexive relation includes, for each
vertex v, the edge (v,v) which connects the vertex to itself. In
other words, *the identity relation is a subset of every reflexive
relation*. The universal relation, of course, is reflexive.
Familiar examples of reflexive relation are:

1. being a blood relative of a person (everybody is his own
relative),

2. being a positive integer divisible by a positive integer
(every positive integer is divisible by itself),

3. being a subset of a set (every set is its own subset).

---

[1]At times, but not in this book, "irreflexive" is used for "antireflexive" and
"asymmetric" for "antisymmetric." Such usage is uncareful.

7.13 SYMMETRIC, ASYMMETRIC AND ANTISYMMETRIC RELATIONS

Another common property of relations is **symmetry**. A relation R is said to be symmetric iff, whenever x is R-related to y then y is also R-related to x.

> **7.11 Definition:** Let A be a set and let R be a relation on A. Then, R is **symmetric** iff, $(x,y) \in R \Rightarrow (y,x) \in R$. A relation which is not symmetric is said to be **asymmetric.**[1]

The reason for the name of the property is easy to discern from Figure 7.11, where each table of a symmetric relation must be symmetric with respect to the *main diagonal* (from left top to right bottom). It must be so because the position of (x,y) is directly across the diagonal and the same distance from it as the position of (y,x). They are as easily recognized from their graphs since, for every directed edge from a vertex v to a vertex u, the edge from u to v must also be there.

R_4	a	b	c
a		x	x
b	x	x	
c	x		

R_5	a	b	c
a	x		
b		x	
c			x

R_6	a	b	c
a			
b			
c			

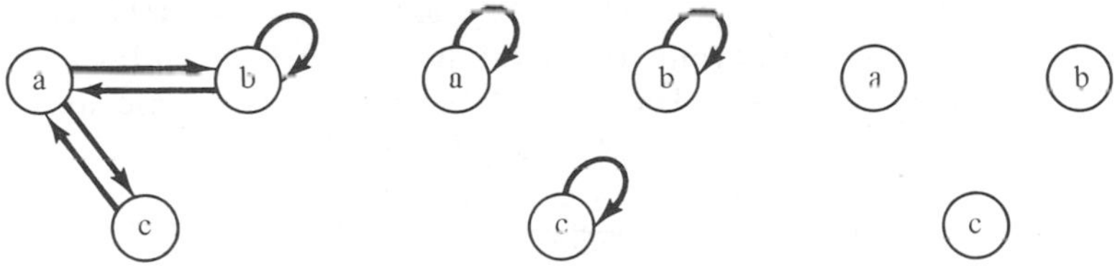

(i)   (ii)   (iii)

Symmetric relations and their graphs

FIGURE 7.11

---

[1]At times, but not in this book, "irreflexive" is used for "antireflexive" and "asymmetric" for "antisymmetric." Such usage is uncareful.

Examples of symmetric relations are:

1. being a blood relative of a person (if I am your relative, so are you mine),
2. being an integer of the same parity as an integer (both even or both odd)
3. being a set which has a common element with a set (if A has a common element with B then B has a common element with A).

The universal, identity, and empty relations are symmetric. As may be evident from the definition and the illustrations, a *relation R is symmetric iff* $R = R^{-1}$.

The opposite extreme of the symmetric relation is one in whose matrix representation no symmetric positions are occupied outside the main diagonal; i.e., if $x \neq y$ then $x\ R\ y$ and $y\ R\ x$ do not both happen. Such a relation is called **antisymmetric.** This property plays an important role in partial orderings, just as the symmetric property is important to equivalence relations.

> *7.12 Definition:*   Let A be a set and let R be a relation on A. Then R is **antisymmetric** iff
>
> $$((x,y) \in R \text{ and } (y,x) \in R) \Rightarrow x = y.$$

Antisymmetric relations are illustrated in Figure 7.12. Note that the empty relation is antisymmetric, as is the identity relation, but the universal relation is not. The only relations that are both symmetric and antisymmetric are subsets of the identity relation. The only relation that is symmetric, antisymmetric and reflexive is the identity relation itself.

In fact a *relation R is antisymmetric iff* $R \cap R^{-1} \subseteq$ identity relation.

The graph of an antisymmetric relation does not possess directed edges in opposite directions between distinct vertices.

Familiar examples of antisymmetric relations are:

1. being the father of a person (the reverse is not true),
2. being an integer greater than (alt., greater than or equal to) an integer,
3. being a subset of a set.

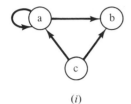 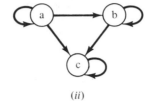 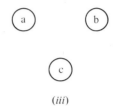

R_7	a	b	c
a	x	x	
b			
c	x	x	

R_8	a	b	c
a	x	x	x
b		x	x
c			x

R_9	a	b	c
a			
b			
c			

(i)   (ii)   (iii)

Antisymmetric relations and their graphs

FIGURE 7.12

7.14 TRANSITIVE AND NONTRANSITIVE RELATIONS

Another frequently encountered property of relations is **transitivity.** As an example, let R be the relation "is a multiple of" on the set of positive integers. Thus, 6 R 2 (or $(6,2) \in R$) since 6 is a multiple of 2, and $(2,6) \notin R$ since 2 is not a multiple of 6. Then R has the property of transitivity, exemplified by: if 12 R 6 and 6 R 3 then it must also be the case that 12 R 3. Here, 6 was the *second* member of the *first* pair and the *first* member of the *second* pair, forming a connection between the first member of the first pair and the second of the second. These two must also be related if the relation is to be transitive.

In general, a relation R on a set A is **transitive** iff, whenever $y \in A$ exists for which there exist x and z in A so that $(x,y) \in R$ and $(y,z) \in R$, then $(x,z) \in R$, for every such x and z.

> *7.13 Definition:* Let A be a set and let R be a relation on A. Then R is **transitive** iff,
>
> $$(x \, R \, y \quad \text{and} \quad y \, R \, z) \Rightarrow (x \, R \, z).$$

A relation which is not transitive is called **nontransitive.**

Figure 7.13 illustrates transitive relations. The reader may note that whether a relation is transitive is not immediately evident from its table illustration. For example, it may not be obvious that, in Figure 7.13 (i), the mark at (b,b) must appear, until one considers the marks at (b,c) and (c,b). As another example, it may not be obvious why the rightmost illustration (part iii) is not transitive until it is compared with part (i).

R_{10}	a	b	c	d
a		x	x	x
b		x	x	
c		x	x	
d		x	x	x

R_{11}	a	b	c	d
a	x	x	x	x
b		x	x	x
c			x	x
d				x

R_{12}	a	b	c	d
a		x	x	x
b		x	x	
c		x		
d		x		x

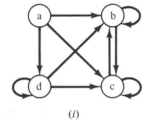

(i)

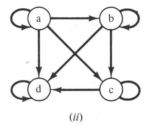

(ii)

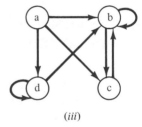

(iii)

Transitive relations   *A nontransitive relation*

FIGURE 7.13

It may be easier to see that R_{12} is not transitive from its graph. For example, since c can be reached from d through b, if R_{12} were transitive, its graph would have an edge from d to c.

7.15 THE TRANSITIVE CLOSURE OF A RELATION

tc(R)

One way to discover whether a relation R is transitive is to construct its **transitive closure** $tc(R)$. (Another common denotation for the transitive closure of R is R^+, similar to the positive closure of a set of strings in Section 4.16 both in idea and in symbol.) Intuitively, if a relation R is not transitive, in order to obtain its transitive closure we must add to the relation such pairs which must be related because of transitivity. This has the effect of inserting such marks into the table representing R which would result in a transitive relation. If R is transitive to begin with, it is identical to its transitive closure. Thus, after obtaining $tc(R)$, the latter may be compared with R; R is transitive iff $R = tc(R)$.

Changing the Boolean matrix of a relation to that of the transitive closure of the relation is accomplished by inserting 1 entries in the appropriate places. A similar change in the graph of the relation amounts to the insertion of directed edges from any vertex to any vertex reachable from it.

The **successor operator,** used in the illustration below, furnishes a convenient means of finding the transitive closure of a relation. (It is not in common use, but provides a more convenient computation than the familiar tools.) We regard the R-relatedness of two elements as "reaching the second from the first" by R. Using as an example the relation R_{13} represented in Figure 7.14, we interpret the mark in row a and column c of the table as, "c **can be reached from** a in one step." In general, $(x,y) \in R_{13}$ iff we can reach y from x in one step. Another, perhaps more convenient description is, "c is an **immediate** $(R_{13}-)$ **successor** of a." Using the former language, for each element we seek to mark all elements eventually **reachable** from it. Using the "successor" terminology, we seek for each element its eventual successors (by any positive number of steps).

R_{13}	a	b	c	d
a			x	
b	x	x		x
c		x		
d		x		

x	Successors of x
a	c
b	a, b, d
c	b
d	b

Table representing R_{13}   *Immediate-successor list of R_{13}*

FIGURE 7.14

We start with the immediate successor list of Figure 7.14, obtained from the table in the obvious way, and we add to it as we find more successors. Thus, since a leads to b, and b leads to d, we should eventually have d (and also b) listed as a successor of a.

7.14 Algorithm (Transitive Closure of a Relation):
Given a relation R on a set A, to find the transitive closure $tc(R)$ of R from the immediate successor list. Let the elements of A be $a_1, a_2, \ldots, a_n$ in the order in which they label the rows of the immediate-successor list (hereinafter called "the list").

Step 1. Initialize $i = 1$. (Do Step 2.)

Step 2. To the left of the list, add a column, called the check column, for a_i and in it place a "$*$" in row i. For every successor a_p of a_i, place a "$-$" in row p of the a_i-check column. (Do Step 3.)

Step 3. If the a_i-check column has no "$-$" (if it has only $+$ marks or blanks), do Step 6. (Otherwise, do Step 4.)

Step 4. Let j be the smallest subscript of any row which has a "$-$" in the a_i-check column. For each k such that a_k appears in the successor column of row j of the a_i-check column, place a "$-$" in row k of the check column if the space is not occupied by a "$-$" or a "$+$". (In case the space is thus occupied, do nothing. Do Step 5.)

Step 5. Cross the "$-$" in row j to get a "$+$". Do Step 3.

Step 6. Enter the label of each row having a "$+$" in its a_i-check column into the successor list in the a_i row, if it is not already there. (Do Step 7.)

Step 7. If $i < n$, increase i by 1 and do Step 2. Otherwise, stop. ∎

The relation R_{13} illustrated in Figure 7.14 has the universal relation as its transitive closure. We illustrate the acquisition of $tc(R_{13})$ in Figures 7.15 — 7.18. In Figure 7.15, we find the eventual successors of a, showing the successive states of the a-check column. In Figure 7.16, the successors of b and c are found. Note that, as soon as every row in the check column is occupied by a "$-$" or a "$+$", the process may be abbreviated since every element is reachable.

Successive States of the
a-Check Column

$+*$	$+*$	$-*$	$-*$	$*$	$*$	$*$	x	Successors of x
$+$	$+$	$+$	$-$	$-$	$-$		a	c
$+$	$+$	$+$	$+$	$+$	$-$	$-$	b	a, b, d
$+$	$-$	$-$	$-$				c	b
							d	b

Finding successors of a

FIGURE 7.15

+	−	x	Successors of x
−*	−*	a	a, b, c, d
−		b	a, b, d
−		c	b
		d	b

(i)

	−	x	Successors of x
+	−	a	a, b, c, d
−*	*	b	a, b, c, d
−		c	b
		d	b

(ii)

Finding successors of b and c

FIGURE 7.16

−		x	Successors of x
+	−	a	a, b, c, d
−		b	a, b, c, d
−*	*	c	a, b, c, d
		d	b

x	Successors of x
a	a, b, c, d
b	a, b, c, d
c	a, b, c, d
d	a, b, c, d

(i) Finding successors of d   (ii) Successor list for tc(R_{13})

FIGURE 7.17

tc(R_{13})	a	b	c	d
a	x	x	x	x
b	x	x	x	x
c	x	x	x	x
d	x	x	x	x

Table representing tc(R_{13})

FIGURE 7.18

In Figure 7.17, the successors of d are found and the successor list for tc(R_{13}) is presented, and in Figure 7.18 the matrix representing tc(R_{13}) shows that the transitive closure of R_{13} is the universal relation.

A relation whose transitive closure is not the universal relation is illustrated in Figure 7.19, with the final states of its check columns shown in part (ii). Part (iii) displays the transitive closure of R_{14}, where the circled marks are the original ones of R_{14}.

It should be clear at this point that, when this algorithm is applied to a transitive relation, the result would be identical to the original relation. The reader may use Figure 7.19 (iii) to verify this statement.

R_{14}	a	b	c	d
a				x
b			x	
c	x			
d				x

Check Columns				x	Successors of x
d	c	b	a		
	+	+	*	a	d
	*			b	c
*		+		c	a
+*	+	+	+	d	d

$tc(R_{14})$	a	b	c	d
a				ⓧ
b	x		ⓧ	x
c	ⓧ			x
d				ⓧ

(*i*) Representing R_{14}　　　　(*ii*) Finding $tc(R_{14})$　　　　(*iii*) Representing $tc(R_{14})$

FIGURE 7.19

To reiterate a statement made earlier, if it is desired to test a relation for transitivity, its transitive closure may be found. The latter is identical with the relation iff the relation is transitive. In Figure 7.19(iii), the original marks from the table of R_{14} are circled. Since some of the marks in the table are not circled, $R_{14} \neq tc(R_{14})$.

The technique just illustrated for finding the transitive closure of a relation is not standard, as was remarked earlier. However, it is considerably faster and more convenient, especially for hard computation, than the customary methods. The most familiar method is that of computing successive **Boolean powers** of the adjacency matrix (the Boolean matrix) of the relation. (Matrix multiplication is illustrated in Section 6.14 and Boolean matrix multiplication in Section 7.7, above.)

Staying with R_{14} of the example of Figure 7.19, above, its Boolean matrix B is

$$B = \begin{bmatrix} 0 & 0 & 0 & 1 \\ 0 & 0 & 1 & 0 \\ 1 & 0 & 0 & 0 \\ 0 & 0 & 0 & 1 \end{bmatrix}.$$

Its second and third Boolean powers (Boolean products of B with itself) are

$$B^2 = B \cdot B = \begin{bmatrix} 0 & 0 & 0 & 1 \\ 0 & 0 & 1 & 0 \\ 1 & 0 & 0 & 0 \\ 0 & 0 & 0 & 1 \end{bmatrix} \cdot \begin{bmatrix} 0 & 0 & 0 & 1 \\ 0 & 0 & 1 & 0 \\ 1 & 0 & 0 & 0 \\ 0 & 0 & 0 & 1 \end{bmatrix}$$

$$= \begin{bmatrix} 0 & 0 & 0 & 1 \\ 1 & 0 & 0 & 0 \\ 0 & 0 & 0 & 1 \\ 0 & 0 & 0 & 1 \end{bmatrix},$$

$$B^3 = B^2 \cdot B = \begin{bmatrix} 0 & 0 & 0 & 1 \\ 1 & 0 & 0 & 0 \\ 0 & 0 & 0 & 1 \\ 0 & 0 & 0 & 1 \end{bmatrix} \cdot \begin{bmatrix} 0 & 0 & 0 & 1 \\ 0 & 0 & 1 & 0 \\ 1 & 0 & 0 & 0 \\ 0 & 0 & 0 & 1 \end{bmatrix}$$

$$= \begin{bmatrix} 0 & 0 & 0 & 1 \\ 0 & 0 & 0 & 1 \\ 0 & 0 & 0 & 1 \\ 0 & 0 & 0 & 1 \end{bmatrix}.$$

Since the set on which the relation is defined has only four elements, the largest Boolean power needed is 3, i.e. one less than the size of the set (Theorem 6.25 in Section 6.9, applied to the graph of the relation). It then remains to form the Boolean sum of the three matrices:

$$B \mid B^2 \mid B^3 = \begin{bmatrix} 0 & 0 & 0 & 1 \\ 0 & 0 & 1 & 0 \\ 1 & 0 & 0 & 0 \\ 0 & 0 & 0 & 1 \end{bmatrix} + \begin{bmatrix} 0 & 0 & 0 & 1 \\ 1 & 0 & 0 & 0 \\ 0 & 0 & 0 & 1 \\ 0 & 0 & 0 & 1 \end{bmatrix}$$

$$+ \begin{bmatrix} 0 & 0 & 0 & 1 \\ 0 & 0 & 0 & 1 \\ 0 & 0 & 0 & 1 \\ 0 & 0 & 0 & 1 \end{bmatrix} = \begin{bmatrix} 0 & 0 & 0 & 1 \\ 1 & 0 & 1 & 1 \\ 1 & 0 & 0 & 1 \\ 0 & 0 & 0 & 1 \end{bmatrix}.$$

The result matrix corresponds to the table in Figure 7.19(iii), but the latter, using Algorithm 7.14, takes considerably less effort to compute. There does exist an algorithm which still uses matrix multiplication, but which represents an improvement in efficiency over the method just illustrated; it is known as *Warshall's algorithm.* However, such improvement is insufficient to justify presenting the algorithm here; it is still far less efficient than Algorithm 7.14, above.

7.16 EQUIVALENCE RELATIONS AND EQUIVALENCE CLASSES

It often comes as a shock to the inexperienced mathematician that **equality** and the two horizontal lines "=" used to denote it require specification, or definition, and that they are not always obvious. One often assumes that "equality" means "identity" and is convinced that "that's the way I've been using it." Isn't 3×4 "the same thing as" 12?

Indeed, it may be necessary to abandon all familiar contexts before this concept is clearly understood. Nonetheless, we first make an attempt to clarify the situation using a common area. In simple arithmetic $2 + 12 = 14$ is true but $2 + 12 = 2$ is false, yet for most nonmilitary personnel 2 o'clock is a sufficient designation for both 2 a.m. and 2 p.m. Even for the purist, $2 + 24$ usually means 2 and not 26 when "telling time." One might argue the differences between analogy and reality, just as one might argue that two instances of "2" are not identical (different ink particles, different geometric-geographical-visual-auditory composition, temporal differences, and the like).

The point of the illustration is that what is "equal" for one purpose may not be for another, and therefore it must be clear for each context what *agreement* is made concerning the meaning of equality.

Two general rules are helpful: 1) every equality must contain identity (i.e. a thing must always equal itself); and 2) *we can allow those things to be equal, among which we do not wish to distinguish.*

Thus, in the collection of *expressions* using the notation of set algebra, $A \cup (A' \cap B)$ and $A \cup B$ are not equal, are *different expressions,* since they are not identical, but in *set algebra* the two are equal since "they have equal value," i.e. an element of either set must be a member of the other, *and that is how equality of sets is defined.*

If we are interested, for some reason, in divisibility of positive integers by 3, we might regard 3, 6, 9, 12, etc. as "the same thing," to wit, belonging to the class of positive integers divisible by 3; any one of them would do just as well as another. We might also regard 1 and 2 as "the same thing," as neither is divisible by 3. But, if we were to further manipulate these classes of positive integers, we might notice that adding any member of $\{1,4,7,10, \ldots\}$ and any member of $\{2,5,8,11 \ldots\}$ results in a number divisible by 3, yet adding two numbers from the same class does not. We then may wish to distinguish between numbers which leave a remainder of 2 after division by 3 and those which leave a remainder of 1, but not among numbers which leave the same remainder. The common notation describing this is similar to the symbol for equality: $1 = 4$ (mod 3), $2 = 5$ (mod 3), $3 = 6$ (mod 3).

There are very many examples on which the reader may draw when thinking of equality of items as their nondistinguishability. All of them have the property that the set of items under consideration is divided into *mutually disjoint classes,* and that items within the same class are not distinguishable from each other but items from distinct classes are always distinguishable.

The division, or partition, of the set A of elements into mutually disjoint classes of nondistinguishable elements is basic to the concept of equality (or **equivalence,** as we shall call it). Each such class is called an **equivalence class** and equivalence (or equality) may be expressed as the relation R by $x \, R \, y$ iff x and y belong to the same equivalence class.

It is now easy to verify that such an **equivalence relation** must be *reflexive, symmetric*, and *transitive*. For, $x \, R \, x$ (read "x is in the same equivalence class as x"), if $x \, R \, y$ then $y \, R \, x$ (read "if x is in the same equivalence class as y, then y is in the same equivalence class as x"), and if $x \, R \, y$ and $y \, R \, z$ then $x \, R \, z$, for all x, y, and z.

In fact, it is easy to create an equivalence relation on a set by simply **partitioning** (subdividing) the set into disjoint classes and defining the relation, as above, by relating all and only members of the same class.

Frequently, equivalence relation is defined in terms of the three properties, and equivalence classes are subsequently defined. Other properties of equivalence relations are then derived, as we do here.

7.15 Definition: Let A be a nonempty set and let R be a relation on A. Then R is an **equivalence relation** iff R is reflexive, symmetric and transitive.

A subset B of A is an **equivalence class** of R (an *R*-**equivalence class**) iff there exists $x \in B$ such that $B = \{y \in A: x \, R \, y\}$.

Thus, an R-equivalence class is a maximal nonempty subset of A in which all elements are R-related.

7.16 Theorem: Let A be a set and let R be an equivalence relation on A. Then the family (set) of all R-equivalence classes is disjoint; moreover, $x \, R \, y$ iff x and y belong to the same equivalence class.

It should be fairly easy to see then that the set of all pairs (x,y) with x and y in an equivalence class B is simply $B \times B$, which leads to a clear and concise formulation of the theorem.

7.17 Theorem: A relation R on a set A is an equivalence relation iff there is a disjoint family A such that

$$R = \bigcup \{B \times B: B \in A\}.$$

Examples of equivalence relations are:

1. The identity relation on a set.
2. The universal relation on a set (all elements are in one block, thus satisfying all three properties).
3. Living on the same floor of an apartment house.
4. Having the same middle initial.

7.17 PARTITIONS

It might be instructive to start with a set (of points labeled for ease of recognition), partition the set, and then construct the equivalence relation induced by the partition, as we do in Figure 7.20. Any other subdivision like the one in part (ii) would yield an equivalence relation, although a different one.

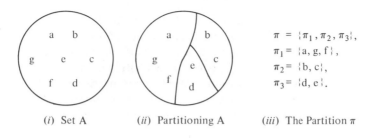

$$\pi = \{\pi_1, \pi_2, \pi_3\},$$
$$\pi_1 = \{a, g, f\},$$
$$\pi_2 = \{b, c\},$$
$$\pi_3 = \{d, e\}.$$

| (*i*) Set A | (*ii*) Partitioning A | (*iii*) The Partition π |

R(π) = {(a, a), (a, g), (a, f), (g, a), (g, g), (g, f), (f, a), (f, g),

(f, f), (b,b), (b, c), (c, b), (c, c), (d, d), (d, e), (e, d), (e, e)}

(*iv*) The Equivalence Relation R(π)

R(π)	a	b	c	d	e	f	g
a	x					x	x
b		x	x				
c		x	x				
d				x	x		
e				x	x		
f	x					x	x
g	x					x	x

(*v*) The Table Representation of R(π)

FIGURE 7.20

The subdivision of a set into mutually disjoint classes is called a **partition.** The concept occupies a prominent position in mathematics and computer science.

7.18 Definition: A **partition** of a nonempty set A is a disjoint family π of nonempty subsets of A whose union is A. In symbols, let $\pi = \{A_i: i \in I\}$ for some nonempty indexing set I, where each A_i is a nonempty subset of A. Then π is a partition of A iff

(i) $A_i \cap A_j = \phi$ iff $i \neq j$;
(ii) $\bigcup_{i \in I} A_i = A$.
Each $A_i \in \pi$ is called a **block** of the partition π.

It is evident then that an equivalence relation on a set A induces a partition on A (the blocks of the partition are the equivalence classes) and that a partition of a nonempty set A induces an equivalence relation on A (two elements are related iff they belong to the same block of the partition).

(Section 7.19, below, deals with partitions with the substitution property, abbreviated as s.p. partitions. More on partitions, s.p. partitions and their relationship with congruence relations and homomorphisms may be found in Chapter 13.)

7.18 REFINEMENT AND COARSENING

Using the example of Figure 7.20, we may further subdivide A by, say, separating a from g and f and e from d, as illustrated by the broken lines in Figure 7.21 (i). The partition π' which results is called a **refinement** of π; it is shown in part (ii) and is represented by a matrix in part (iii).

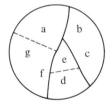

$$\pi' = \{\pi_{11}, \pi_{12}, \pi_2, \pi_{31}, \pi_{32}\},$$
$$\pi_{11} = \{a\}, \ \pi_{31} = \{d\},$$
$$\pi_{12} = \{g, f\}, \ \pi_{32} = \{e\},$$
$$\pi_2 = \{b, c\},$$

(*i*) Further Subdividing A   (*ii*) The Resulting Refinement

	a	b	c	d	e	f	g
a	x						
b		x	x				
c		x	x				
d				x			
e					x		
f						x	x
g						x	x

(*iii*) The Representing Table

FIGURE 7.21

We may obtain a proper coarsening π'' of π of Figure 7.19 by combining classes, as is shown in Figure 7.22.

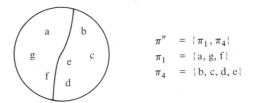

$$\pi'' = \{\pi_1, \pi_4\}$$
$$\pi_1 = \{a, g, f\}$$
$$\pi_4 = \{b, c, d, e\}$$

(*i*) Combining Blocks of π (*ii*) The Resulting Coarsening

	a	b	c	d	e	f	g
a	x					x	x
b		x	x	x	x		
c		x	x	x	x		
d		x	x	x	x		
e		x	x	x	x		
f	x					x	x
g	x					x	x

(*iii*) The Representing Table

FIGURE 7.22

7.19 Definition: Let π and ρ be two partitions of a set A. Then ρ is a **refinement** of π iff every block of ρ is a subset of a block of π. Moreover, ρ is a **proper refinement** of π iff it is a refinement of π and at least one block of ρ is a *proper* subset of a block of π. ρ is a refinement (alt., proper refinement) of π iff π is a **coarsening** (alt., **proper coarsening**) of ρ.

It is noteworthy in the preceding examples that the disjointness of the blocks permits easy inspection of reflexivity (the main diagonal is occupied), symmetry (the marks appear in symmetric positions across the main diagonal), and transitivity (each block is full of marks and there is no contact point between blocks). This serves to intuitively reinforce the conclusion that the relation induced by a partition, relating members of the same block, is an equivalence relation.

7.19 PARTITIONS WITH THE SUBSTITUTION PROPERTY

This topic is more appropriate to an algebraic setting and is, indeed, discussed in greater detail in Section 13.7. Here, we only present and illustrate the basic concept.

Consider as an example the set $\mathbb{P}$ of all positive integers, with special attention to the operations of addition and multiplication. Let π be the partition of $\mathbb{P}$ into the three equivalence classes $A = \{3,6,9,12,\ldots\}$, $B = \{1,4,7,10,\ldots\}$ and $C = \{2,5,8,11,\ldots\}$. Now, $4 + 5 = 9$ is a member of A, but the same will be the case if we *substituted* any member of B for 4 and any member of C for 5: $7 + 8 = 15$ is also a member of A. In fact, we could make a symbolic addition table *for the blocks of the partition*, because they **preserve** the addition on $\mathbb{P}$:

$+$	A	B	C
A	A	B	C
B	B	C	A
C	C	A	B

As is easily seen, π preserves multiplication on $\mathbb{P}$, as well.

When the substitution of *any* representative of the same block of the partition does not change the block in which the result resides, we say that the partition has the **substitution property,** and that it is a **substitution-property partition** (abbreviated as **s. p.-partition**).

By way of contrast, consider the function f on the set of positive integers, defined by

$$f(p) = \begin{cases} 1, \text{ if } p \text{ is prime,} \\ 0, \text{ otherwise.} \end{cases}$$

The partition π does *not* preserve the function f, since both 7 and 4 are in B, both 5 and 8 in C, etc.

7.20 RESTRICTIONS AND EXTENSIONS OF RELATIONS

In the special relations we call functions (Chapter 9), we are often interested in the **restriction** of the function to a subset of its domain, that is, its behavior on *some* of the members of the domain; frequently, this is done because the function possesses special properties on the subdomain. The same is true in the general case of relations and we introduce the concept here in the more general setting.

Consider the relation R on the set $A = \{a,b,c,d,e\}$ represented in Figure 7.23 (i). R is not an equivalence relation for such reasons as $(c,c) \notin R$, violating reflexivity; $(c,a) \in R$ but $(a,c) \notin R$, violating symmetry; and $(e,a) \in R$ and $(a,b) \in R$ but $(e,b) \notin R$, violating transitivity. However, when we restrict R to the subdomain $B = \{a,b,d\}$, the result is an equivalence relation on the subdomain, as is seen in part (ii) of Figure 7.23. In essence, we intersected the relation R, which is the set of ordered pairs of related elements, with the Cartesian product $B \times B$, to obtain the restriction of R to B.

R	a	b	c	d	e
a	x	x			
b	x	x	x		
c	x				
d			x	x	
e	x				x

	a	b	d
a	x	x	
b	x	x	
d			x

(*i*) Representing R   (*ii*) R Restricted to {a, b, d}

FIGURE 7.23

7.20 Definition: Let A be a set, let R be a relation on A and let $B \subseteq A$. Then $R \cap (B \times B)$ is the **restriction of R to B**.

The notation often used to denote the restriction of R to B is $R|_{B \times B}$ or $R|_B$ (the latter is more common with functions).

In a similar vein, we may *extend* a relation to a larger set than its domain by defining a relation on the larger set which is identical to the original relation on the original domain. This permits the inclusion of all, some, or none of the new pairs into the newly defined relation.

7.21 Definition: Let A and B be sets, let $A \subseteq B$, let R be a relation on A, and let T be a relation on B. Then T is an **extension of R to B** iff R is a restriction of T to A.

8

Orderings

8.1 PARTIAL ORDERING

Equivalence relations take no heed of any order of precedence; thus, when two elements of a set are equivalent they have equal "ranking" and when they are not, neither is recognized as preceding the other in order.

We often wish to recognize precedence, such as when comparing the sizes of real numbers (less than, less than or equal to), noting which set is a subset of a set, and the like. Using the subset relation as an example in which precedence may be important, we find that the relation is reflexive since every set is its own subset, that is, $A \subseteq A$. Similarly, if $A \subseteq B$ and $B \subseteq C$ then $A \subseteq C$, and thus the subset relation is transitive. The difference between such a relation and an equivalence relation must be in the requirement for symmetry. In fact, the negation of symmetry, i.e. asymmetry, is not sufficient and what is required is *antisymmetry,* so that $A \subseteq B$ and $B \subseteq A$ together imply $A = B$. If this were not the case, a basic property of sets would be missing, to wit, two sets are equal iff each is a subset of the other.

A relation on a set with these three properties, reflexive, antisymmetric, and transitive, is called a **partial ordering** on the set. The reason for the word "partial" is that not every pair of elements of the set must be related. This is seen when we consider the set of all subsets of a set A, i.e. $\mathcal{P}(A)$. Two such subsets may exist with neither a subset of the other, as is shown in Figure 8.1. We have $C \subseteq A$, but neither $A \subseteq B$ nor $B \subseteq A$. In such a case, A and B are said to be **not comparable** or **incomparable**.

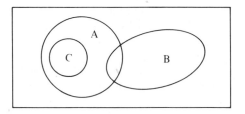

A Venn diagram

FIGURE 8.1

Probably the most familiar example of partial ordering is the "less than or equal to" relation defined on real numbers, where "$x \leq y$" is read "x is less than or equal to y." This is the reason that the symbol "$\leq$" is often used to define a partial ordering, even when the interpretation is different from "less than or equal to." The reader should be cautious not to ascribe to the symbol a meaning other than the intended one, or in addition to it.

8.1 Definition: Let A be a set and let R be a relation on A. Then, R is a **partial ordering** on A iff R is

(i) reflexive ($\forall\ x \in A, x\ R\ x$),

(ii) antisymmetric (if $x\ R\ y$ and $y\ R\ x$ then $x = y$), and

(iii) transitive (if $x\ R\ y$ and $y\ R\ z$ then $x\ R\ z$).

R may be denoted by $\leq$ and "$x \leq y$" may be read "x precedes y," "x is in y," "x is a part of y," "x is contained in y."

The antisymmetric property of partial ordering makes it necessary to have "equality" defined on the set or, in other words, an equivalence relation selected to be used in defining the partial ordering. However, that is usually implicit in the description of the partial ordering, as may be seen from the following examples.

Example 1:
The relation "divides" (x and y) on the set of positive integers. (Equality of positive integers is used.)

Example 2:
On the set of single-valued functions over the interal [0,1] (inclusive of 0 and 1), the relation "$\leq$" defined by: $\forall f, g, f \leq g$ iff

$f(x) \le g(x)$ for every x in the interval. (Equality of functions is taken as the equivalence relation, where $f = g$ iff $f(x) = g(x)$ for every x in $[0,1]$).

It is interesting to note in the second example that, had we chosen the equivalence relation otherwise, say

$$f = g \text{ iff } |f(x) - g(x)| < 1, \forall x \in [0,1],$$

that is f and g are never as far apart as a unit, then the interpretation of $f \le g$ would be quite different. For instance, with the two constant functions $f = 3$ and $g = 2.5$ over $[0,1]$, we still have $f \le g$ (Figure 8.2 (i)). Even more unorthodox may be the example in Figure 8.2 (ii), where f and g are defined as the linear functions $f(x) = x$ and $g(x) = -2x + 2$ over $[0,1]$. Hereinafter, mainly so that equality be reflected by the main diagonal of the representing table, we shall regard the equality in the antisymmetric property as identity, on the assumption that the displayed elements are in truth equivalence classes of the original elements.

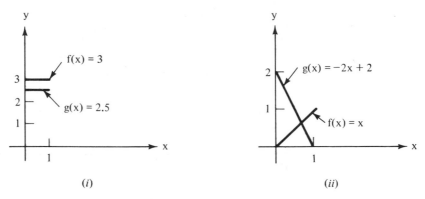

Unusual interpretations of $f \le g$

FIGURE 8.2

8.2 PARTIALLY ORDERED SETS (POSETS)

A set A with a partial ordering $\le$ is called a **partially ordered set** or **poset** (for short) and is denoted as the pair $(A, \le)$.

The **sense** of the partial ordering is often reversed for

$\geqq$

convenience, so that "$x \leqq y$" means the same as "$y \geqq x$" and the appropriate adjustments are made in the terminology (greater than, includes, is a superset of, etc.).

By the definition of partial ordering, it is clear that the main diagonal of the representing table must be entirely marked, because of the reflexivity requirement. (See Figure 8.3.) When it is desired to refer to the relatedness of a pair off the main diagonal, for instance (a,b) in Figure 8.3, and when the emphasis is on the pair being *off* the main diagonal, we

<

have $a \leqq b$ and $a \neq b$ and we may write $a \lneqq b$ or $a < b$ and say that a is *less than b, properly contained* in b, and the like. We

>

may similarly use $b > a$.

$\leqq$	a	b	c	d
a	x	x	x	
b		x	x	
c			x	
d	x	x	x	x

Representing a partial ordering

FIGURE 8.3

The relation "$<$" just discussed is called by some a *nonreflexive partial ordering* and has the following properties. (Essentially, we may derive a strict ordering from a partial one, and vice versa.)

8.2 Theorem: Given a poset $(A,\leqq)$. Then,

(i) for no $a \in A$, $a < a$;

(ii) if $a < b$ and $b < c$ then $a < c$.

8.3 Theorem: Let A be a set, let R be a relation on A which satisfies (i) and (ii) of Theorem 8.2, and let "$=$" be the identity relation on A. Let $a \leqq b$ mean $a R b$ or $a = b$. Then the relation $\leqq$ is a partial ordering on A.

8.3 DUALITY OF POSETS

In Section 4.2, above, in discussing the principle of duality, it was remarked that if set inclusion appears in the statement whose dual is desired, the sense of the inclusion should be reversed; that is, the dual of $A \subseteq B$ is $B \subseteq A$. This derives from the more general **principle of duality for posets:**

(i) The inverse of any partial ordering is itself a partial ordering.

(ii) The **dual of a poset** $(A, \leq)$ is the poset $(A, \geq)$, where "$\geq$" is the inverse of "$\leq$."

(iii) Posets are dual in pairs, when they are not self-dual.

(iv) Definitions and theorems on posets are dual in pairs when they are not self-dual.

This principle of duality applies to algebra, Boolean algebra, logic, and projective geometry as well.

8.4 ORDERING

As may have been intuitively concluded by now, the essence of *ordering* a set of elements is the transitive property, for it establishes the "order of precedence." Indeed, the two terms are used synonymously.

> *8.4 Definition:*   A relation R on a set A is an **ordering** of A iff it is a transitive relation on A.

Clearly, then, this gives rise to orderings which are not very interesting. For example, $A \times A$ is an ordering of A. Further conditions on the relation result in more restrictive, and thus more interesting, orderings. Antisymmetry is one such condition and it helped produce partial orderings. These, however, are not forced to relate every pair of distinct elements of the domain, and thus they allow elements which are not comparable.

8.5 LINEAR ORDERING

Probably the most familiar example of **linear ordering** (also **total, complete, simple ordering**) is the "less than" relation defined on the real numbers, where "$x < y$" is read "x is less than y." (To determine when two real numbers x and y are related in this fashion, see if their difference is positive; that is, $x < y$ iff $y - x$ is positive.)

As the word "linear" intimates, the elements in a **linearly ordered set** may be thought of as arranged on a (drawn) line, where "x is to the left of y" on a horizontally drawn line means $x < y$. This visualization points out the important feature of a linear ordering:

Every two distinct elements are related.

That is, if x and y are distinct elements of a linearly ordered set, then either $x < y$ or $y < x$. This property is called **dichotomy.**

The reader must be cautioned here of a fairly common error in texts of elementary mathematics. A linear ordering is often defined as a partial ordering which is also dichotomous. This is false, because *a linear ordering does not have to be reflexive.*

More specifically, the following "definition" is often seen:

A partial ordering $\leq$ on a set A is a linear ordering on A iff, for every $a,b \in A$, either $a \leq b$ or $b \leq a$.

This is *not a definition* of linear ordering; it just tells when a *partial* ordering is a linear ordering. However, it often gives rise to the *mistaken* notion that a linear ordering is (always) a special kind of partial ordering. The relation "less than" on the set of real numbers is a counterexample: it is regarded by all as a linear ordering but it is not reflexive and therefore is not a partial ordering.

It is true, of course, that the partial ordering "less than or equal to" ($\leq$) on the real numbers is at the same time a linear ordering of the real numbers. It underscores the **trichotomy:**

For any two real numbers x and y, exactly one of the following is true: $x < y$, $x = y$, $y < x$.

8.5 Definition: Let A be a set and let R be a relation on A. Then R is a **linear ordering** of (or on) A iff R is

(i) *antisymmetric* (if $x R y$ and $y R x$ then $x = y$),

(ii) *transitive* (if $x R y$ and $y R z$ then $x R z$), and

(iii) *dichotomous* (if $x \neq y$, then either $x R y$ or $y R x$).

The set A together with the linear ordering R is called a **linearly ordered set,** also a **chain.** A is then said to be **linearly**

ordered by R.

It is easy to construct new examples of a certain ordering from the examples given previously. Take a subset of the given set and use the same ordering, but restrict it to the subset. More precisely,

> **8.6 Theorem:** Any subset B of a poset A is itself a poset, under the same partial ordering restricted to B. Likewise, any subset of a chain is itself a chain (under the restricted linear ordering).

To summarize, the relations $\leq$ and $<$ on the set of real numbers $\mathbb{R}$ are linear orderings of $\mathbb{R}$. The relation $\leq$ is reflexive while $<$ is not. The relation $<$ is antisymmetric by default, i.e. it is not possible for any two real numbers x and y to have both $x < y$ and $y < x$. And it is true for both relations that, given any two distinct real numbers x and y, either both $x \leq y$ and $x < y$ or both $y \leq x$ and $y < x$.

8.6 LEXICOGRAPHIC ORDERING

Another interesting linear ordering is the one called **lexicographic,** as it is used in alphabetizing lists such as lexicons, dictionaries, telephone and address books, and the like. It may be applied to any set whose members are sequences (finite or infinite) of elements, and there is a linear ordering on these component elements. In this fashion, the word *cape* precedes the word *catch* in the dictionary since the first position in which the words differ is the third, and the third component p of cape precedes the third component t of catch in the alphabetical ordering of the single third components. For the same reason, the sequence 3243 of digits should precede 327 in the lexicographic ordering, assuming the usual ordering of the digits since $4 < 7$, even though the numerical value of 327 is smaller than that of 3243.

> **8.7 Definition:** Let A be a set of sequences of the form $x = (x_1, x_2, \ldots)$ and let a linear ordering $<_i$ be defined on each i-component of the members of A, for each position i in the sequence. Then a sequence $a = (a_1, a_2, \ldots)$ precedes a sequence $b = (b_1, b_2, \ldots)$ in the **lexicographic ordering** of A iff, for the smallest integer p such that $a_p \neq b_p$, it is true that $a_p <_p b_p$.

The lexicographic ordering is often used for linearly ordering sets which are not usually linearly ordered, such as the set of complex numbers: $a + ci < b + di$ iff either $a < b$, or, if $a = b$ then $c < d$.

8.7 LATTICES

Before leaving the topic of orderings and partial orderings in particular, the concept of **lattice** has to be explored as an added structure to partial ordering. Here, again, diagrams are an aid in visualizing the concepts and relationships involved. A poset $(A, \leqq)$ lends itself well to a diagram, provided the set A is not too large. We adopt the convention that, of two elements connected with a line segment, or a chain of line segments, the lower precedes ($\leqq$) the higher. (Thus the diagram is that of a directed graph, but not necessarily that of a tree.)

$\leqq$	a	b	c	d	e
a	x	x	x		x
b		x			
c			x		x
d			x	x	x
e					x

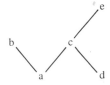

(*i*) A Partial Ordering, $\leqq$ (*ii*) A Diagram for $\leqq$

FIGURE 8.4

A moment's inspection should suffice to show how the two parts of Figure 8.4 are related. In fact, any diagram of the sort that is shown in part (ii), with but some obvious precautions, represents a partial ordering: reflexivity is assumed, antisymmetry is free as long as no element appears twice, and transitivity is achieved by chaining connecting line segments.

Figure 8.5 illustrates divisibility on the positive integers from 2 to 12, while Figure 8.6 shows the partial ordering of set inclusion on the set of subsets of $\{a,b,c\}$.

An element x of a poset $(A, \leqq)$ is a **lower bound** of a subset B of A if $x \leqq y$ for every $y \in B$. Thus, in Figure 8.5, $\{4,6\}$ has 2 as its only lower bound, but $\{4,6,9\}$ has none. In Figure 8.6, any collection of subsets has a lower bound; for example, $\{\{a,b\},\{b,c\}\}$ has $\{b\}$ and ϕ as lower bounds, $\{\{a,b\},\{a,b,c\}\}$ has

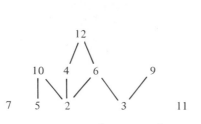

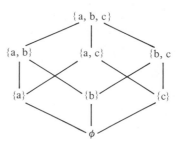

Divisibility on {2,3 . . . , 12}

FIGURE 8.5

$\mathcal{P}(\{a,b,c\})$ *with* $\subseteq$

FIGURE 8.6

$\{a,b\}$, $\{a\}$, $\{b\}$ and ϕ as lower bounds, and the entire set $\{\{a,b,c\}\}$ has ϕ as a lower bound.

An element x of a poset $(A,\leq)$ is an **upper bound** of a subset B of A iff $y \leq x$ for every $y \in B$. In Figure 8.5, $\{2,4\}$ has 4 and 12 as upper bounds, $\{2,3,4,6\}$ has only 12 as an upper bound, and $\{7,10\}$ does not have one. Any collection of subsets in Figure 8.6 has at least one upper bound.

In Figure 8.5, $\{4,12\}$ has 2 and 4 as lower bounds but 4 is the *greatest* lower bound, because $2 \leq 4$ and $4 \leq 4$. An element

glb

x of a poset $(A,\leq)$ is a **greatest lower bound** (abbreviated **glb**) of a subset B of A iff x is a lower bound of B and, for every lower bound y of B, $y \leq x$; i.e., if $y \leq z$ for every $z \in B$, then $y \leq x$.

lub

Similarly, x is a **least upper bound** (**lub**) of B iff it is an upper bound of B and precedes every upper bound of B; i.e., if $z \leq y$ for every $z \in B$, then $x \leq y$.

We leave it to the reader to observe the glb and lub of various subsets of the posets in Figures 8.5 and 8.6. It should be pointed out, however, that it is not always the case that two elements have a lub (alt., glb), as is evident from the many cases of Figure 8.5. On the other hand, once there is a lub for two elements, *there is exactly one;* the same holds true for the glb of two elements—*if one exists, it is unique.*

The glb of two elements x and y is also called the **meet** and

∧
∨

is denoted by $x \wedge y$. The lub of x and y is also called the **join** and is denoted by $x \vee y$.

In Figure 8.6, $\{a\} \wedge \{b\} = \phi$ and $\{a\} \vee \{b\} = \{a,b\}$; $\{a\} \wedge \{b,c\} = \phi$ and $\{a\} \vee \{b,c\} = \{a,b,c\}$; $\{a,b\} \wedge \{b\} = \{b\}$ and $\{a,b\} \vee \{b\} = \{a,b\}$.

The reader will have noticed by now that, in Figure 8.6,

every two members have both a lub and a glb. Such a poset is called a **lattice.**

The meet and the join, or glb and lub, are defined in terms of the partial ordering, but once they are defined in a lattice they may be regarded as binary operations on the elements of the lattice, as the immediately preceding illustrations indicate.

> ***8.8 Definition:*** A **lattice** is a poset, any two of whose elements have a glb and a lub.

The preceding development should be sufficient for most uses of lattices in computer science. For the less frequent encounters, the following are presented without illustration and with scant comment.

> ***8.9 Definition:*** A lattice L is said to be **complete** iff each of its subsets has a lub and glb in L.

It is of interest (and some importance) to note that the poset in Figure 8.5 has several **maximal elements** (which precede no other elements), 7, 10, 12, 9, 11, and several **minimal elements** (which have no other elements preceding them), 7, 5, 2, 3, 11. Yet, the poset in Figure 8.6, which is a lattice, has exactly one maximal element, $\{a,b,c\}$ and one minimal element, ϕ. It is so because *all finite lattices are complete* and in a complete lattice there is one lub and one glb for the entire lattice.

> ***8.10 Theorem:*** The operations glb ($\wedge$) and lub ($\vee$) on a lattice are commutative, associative, and idempotent. (See Theorem 3.13).

> ***8.11 Theorem:*** The principle of duality holds for lattices with the partial ordering and its inverse being a dual pair, as well as $\wedge$ and $\vee$.

> ***8.12 Definition:*** A lattice $(L,\leq,\wedge,\vee)$ is called **distributive** iff its meet and join are distributive over each other; i.e. iff, $x,y,z \in L$,
>
> *(i)* $x \wedge (y \vee z) = (x \wedge y) \vee (x \wedge z)$, and
> *(ii)* $x \vee (y \wedge z) = (x \vee y) \wedge (x \vee z)$.

8.13 Definition: Let L be a lattice, let L have a unique maximal element called 1 and a unique minimal element called 0. Then L is said to be a **complemented** lattice iff, for every $x \in L$ there exists a $y \in L$ (called the **complement** of x) such that

$$x \wedge y = 0 \text{ and } x \vee y = 1.$$

In a complemented distributive lattice, the involution law as well as DeMorgan's laws hold. (See Table 4.1, above).

8.14 Definition: A lattice with 0 and 1 which is distributive and complemented is a **Boolean algebra.**

9

Functions

9.1 A FUNCTION IS A SINGLE VALUED RELATION

In elementary mathematics, the beginner often confuses a function with its diagrammatic presentation and fails to make a distinction between a function f and its *value $f(x)$* on a member x of its domain. The *domain* and *range* of a function are also often disregarded, making two distinct functions appear to be the same just because their values coincide on the common parts of the two domains.

In Figure 9.1 appears part of the diagram of the "squaring function," the one whose value at a real number is the square of the number. The function is often written as $y - f(x) - x^2$, even though this is only the value (height, in the diagram) of the function at a number x. The function may be regarded

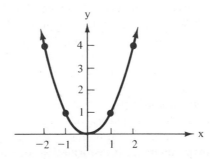

A part of the squaring function

FIGURE 9.1

(with some reservations) as the rule which assigns to each element of the domain, called **argument,** the square of the argument, called the **value** or **image** of the function at the argument. This is (almost) the same as saying that *the function f is the set of ordered pairs* of the form (x, x^2), with every real number appearing as a first coordinate of a pair in the set. Thus, the function illustrated in Figure 9.1 is

$$\{(x, x^2): x \in \mathbb{R}\}.$$

Indeed, a function is a relation, i.e. a set of ordered pairs (see Chapter 7), with the special property that no two different pairs have the same first coordinate. In other words, a function is a *single valued* relation—to each member of its domain there is assigned a unique member of the range.

In the diagrammatic illustration, this means that no two points on the same vertical, i.e. one directly above another, may belong to the illustration. In terms of a table representing a relation, as illustrated in the previous sections, the uniqueness of the assigned value means that every row of the matrix may contain no more than one mark.

As was the case with relations, the domain and range are important parts of the function and must not be overlooked. For instance, "the squaring function" does not quite describe the function, although it provides the rule for assigning a value to a given member of the domain. To illustrate the point, consider the two functions depicted in Figures 9.2 and 9.3. Both are still the "squaring function" which assigns to a number x of the domain the value x^2. Yet the illustrations look quite different from that of Figure 9.1 and from each other. The reason is, of course, that the domains are different, *which makes the functions different.* The domain in Figure 9.1 is the set $\mathbb{R}$ of real numbers, in Figure 9.2 it is the set of nonnegative real numbers, and in Figure 9.3 it is the set of integers.

As was the case with relations then, two functions, f and g, are **equal,** or identical, iff they have the same members (ordered pairs). Clearly this is the case iff f and g *have the same domain* (the set of first coordinates of the ordered pairs) and $f(x) = g(x)$ for each x in this domain.

The term "function" has as synonyms **correspondence, map, mapping, transformation, operator,** and others. We shall use "function" almost exclusively.

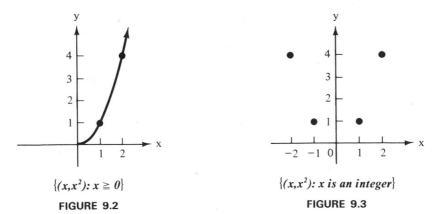

$\{(x,x^2): x \geq 0\}$

FIGURE 9.2

$\{(x,x^2): x \text{ is an integer}\}$

FIGURE 9.3

9.1 Definition: A relation f (a collection of ordered pairs) is a **function** iff, whenever (x,y) and (x,z) are members of f, then $y = z$.

The set of all first coordinates (left members) of the ordered pairs of f is the **domain** of f. The set of all second coordinates is the **range** of f. The second coordinate $f(x)$ of the pair $(x, f(x))$ is called the **value,** or **image,** of f at x and the function f is said to **assign** the value $f(x)$ to x, or to **take the value** $f(x)$ at x. If a set is specified where the function f takes values, such a set is called the **co-domain** of f. The co-domain of f may, but need not, coincide with the range of f.

Two functions f and g are **equal** iff they have the same domain and assign to each member of the domain the same value; i.e., if they have exactly the same ordered pairs.

To illustrate the difference between range and co-domain, the function f illustrated in Figure 9.1 may be specified with domain and co-domain both being the set $\mathbb{R}$ of real numbers (written $f: \mathbb{R} \rightarrow \mathbb{R}$), but f never takes negative values, so that the range of f is the set of nonnegative real numbers.

9.2 POINT FUNCTIONS AND SET FUNCTIONS

Because of the importance of the domain and range in the specification of the function, we write $f: D \rightarrow C$, pronounced "f

is a function on D to C," to indicate that f is a function with domain D and co-domain C. Thus, we would specify the function illustrated in Figure 9.1 as

$f\colon D \to R$

$$f\colon D \to \mathbb{R} \to \mathbb{R}, \text{ such that } f(x) = x^2, \forall x \in \mathbb{R}.$$

The range of f is also called the **set of images** of f, or the f-**image** of the domain D, and is denoted by

$$f(D) = \{c \in C\colon c = f(d), \text{ for some } d \in D\}.$$

This notation and notion are quite important, since a function whose arguments are members of the domain D, called a *point function,* is thereby converted to a function defined on a *set* of points, whose argument and whose value are *sets* rather than points; such a function is called a **set function.** We return to this idea in more generality later.

9.3 EPIC (ONTO) FUNCTIONS

To reiterate, the only way a function is more restricted than a relation is the uniqueness of the value for each member of the domain: where a relation may have any number of distinct ordered pairs with the same first coordinate, a function can have only one. This restriction may not seem very consequential at first glance, but it turns out to be of considerable importance; functions are used much more frequently and in many more contexts than relations which are not functions. Furthermore, special functions and functions with special properties play important roles in all of mathematics and its applications.

In general, there is no restriction on the number of ordered pairs in a function with the same right coordinate; that is, a function in general is "many to one" in the sense that there may be many arguments at which the function has the same value. For example, the function $f\colon \mathbb{R} \to \mathbb{R}$ with $f(x) = x^2$, $\forall x \in \mathbb{R}$, illustrated in Figure 9.1, is symmetric about the y-axis and thus has the value 1 both at $x = -1$ and at $x = 1$, i.e. $f(-1) = f(1) = 1$; also $f(-2) = f(2) = 4$; etc. An extreme example of this feature is a **constant function,** whose range is a singleton, i.e. every ordered pair in the function has the same (constant) second coordinate. Such a function is $g\colon \mathbb{R} \to \mathbb{R}$ such that $g(x) = 2$, $\forall x \in \mathbb{R}$, illustrated in Figure 9.4.

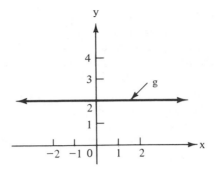

A constant function

FIGURE 9.4

We return to the distinction made in Definition 9.1 between the range and the co-domain of a function. A function $f: D \rightarrow C$ **on** the domain D **to** the co-domain C has its range as a subset of C, i.e. $f(D) \subseteq C$. However, it may happen that f exhausts the co-domain; that is, $\forall y \in C$ there is some (at least one) $x \in D$ such that $f(x) = y$. In other words, the range of f is *all* of the co-domain C.

> **9.2 Definition:** Let $f: D \rightarrow C$ be a function with domain D and co-domain C. Then f is said to be **on** D and **to,** or **into** C. The function f is said to be **onto** C iff $f(D) = C$. In that case, f is called **epic** (also a **surjection**).

Of course, every function is onto its own range; the distinction becomes useful only when the range is not known to equal the co-domain.

9.4 COMPOSITION AND ITERATION OF FUNCTIONS

Composition of relations was defined in Section 7.7, but illustration and further discussion were deferred until this presentation of functions. The essential idea in composition of two functions is that *an image by the first function becomes an argument of the second.* For example, if one function $f: \mathbb{P} \rightarrow \mathbb{P}$ ($\mathbb{P}$ is the set of positive integers) assigns to an argument x its double, i.e. $f(x) = 2x$, and another function $g: \mathbb{P} \rightarrow \mathbb{P}$ assigns to an argument x its square, i.e. $g(x) = x^2$, then the value "f of g" of an argument x, written $(f \circ g)\,(x)$ or $(fg)(x)$, is the double of the square, i.e. $(f \circ g)(x) = 2x^2$, while "g of f" of x is the square of the double, i.e. $(g \circ f)(x) = (2x)^2 = 4x^2$.

As long as the range of the first function f is a subset of the domain of the second g, the value $g(f(x))$ is defined and is taken as the value $(g \circ f)(x)$ of the composition $g \circ f$ at x. This is illustrated in Figure 9.5 for $g \circ f$ and in Figure 9.6 for $f \circ g$, where f is the "doubling" function and g the "squaring" function.

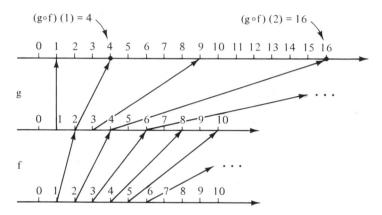

Illustrating the composition $g \circ f$

FIGURE 9.5

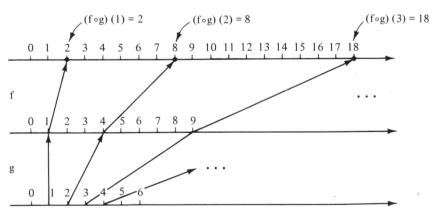

Illustrating the composition $f \circ g$

FIGURE 9.6

9.3 Definition: Let $f: A \to B$ and $g: B \to C$. Then, the **composition** $g \circ f: A \to C$ is the function defined by:

$$\forall x \in A, (g \circ f)(x) = g(f(x)).$$

Obviously, the definition of composition of functions is extendable to more than two functions in the obvious way. Care must be taken though that the domain of a composed function contain the range of its predecessor.

A frequent case of composition of functions occurs when a function has the same domain and co-domain and the composition is of the function with itself, however many times. In such cases the resulting composition is called the **power** of the function and the notation uses exponentiation.

Let $h: \mathbb{P} \to \mathbb{P}$ be defined as follows:

$$h(x) = \begin{cases} \dfrac{x}{2}, & \text{if } x \text{ is even,} \\[2ex] \dfrac{x+1}{2}, & \text{if } x \text{ is odd.} \end{cases}$$

Then a list of the "first" ten ordered pairs of h appears in Table 9.1. Now, composing h with itself yields $h^2 = h \circ h$, whose explicit definition is slightly involved, but whose images are easy to compute by **iterating** h. Thus, $h^2(5) = (h \circ h)(5) = h(h(5)) = h(3) = 2$. Similarly, we could obtain the value of h^3 at 10 as follows: $h^3(10) = (h \circ h \circ h)(10) = h(h(h(10))) = h(h(5)) = h(3) = 2$.

	TABLE 9.1		
	ILLUSTRATING h		

x	$h(x)$	x	$h(x)$
1	1	6	3
2	1	7	4
3	2	8	4
4	2	9	5
5	3	10	5

TABLE 9.2

ILLUSTRATING POWERS OF h

x	$h(x)$	$h^2(x)$	$h^3(x)$
1	1	1	1
2	1	1	1
3	2	1	1
4	2	1	1
5	3	2	1
6	3	2	1
7	4	2	1
8	4	2	1
9	5	3	2
10	5	3	2

It should be noted that *composition* as a binary operation on functions *is associative,* a fact which we used in the previous illustration.

9.4 Theorem: Let $f: A \rightarrow B$, $g: B \rightarrow C$, and $h: C \rightarrow D$ be functions. Then, $(h \circ g) \circ f = h \circ (g \circ f)$.

TABLE 9.3

DEFINING $k: D \rightarrow D$ AND ITS POWERS

x	$k(x)$	$k^2(x)$	$k^3(x) = k(x)$
a	c	a	c
b	d	d	d
c	a	c	a
d	d	d	d

We illustrate the same concept of powers of a function but this time on a finite set and, at the same time, stress the idea that a function need not represent a *familiar* situation. As was mentioned earlier, a function may be defined by specifying its domain and then prescribing the assignment of an image to each member of the domain. In the present illustration we do precisely that: we let the domain be $D = \{a,b,c,d\}$ and we define the function $k: D \rightarrow D$ by Table 9.3, which details the pairing of arguments with their values. (The domain and range need not be the same, although they are so in the example.)

The table stops with k^3 because $k^3 = k$ (the images by the two functions are identical). The phenomenon encountered here, where a power of the function equals a smaller power of the same function, is one which cannot be avoided for finite domain and range. (In fact, if there were a choice in the matter, the effect would be desirable, because no further powers need be displayed: if $k^3 = k$ then $k^4 = k^2$, $k^5 = k^3$, etc.)

The reason is that when both domain and range are finite, the image of each domain member can be any, but only, one of the members of the range, and there are only finitely many of those. Thus, if the range has n members, the image of each argument is one of n possibilities and therefore there are only n different ways to assign it. If the domain has m members, then the previous sentence holds true for each of them; consequently, the images of the first two arguments can be assigned in $n \cdot n = n^2$ ways, those of the first three arguments in $n \cdot n \cdot n = n^3$ ways, and the images of all m arguments can be assigned in n^m different ways.

9.5 THE SET OF ALL FUNCTIONS ON
A TO *B*

As an example consider the domain $A = \{a,b,c\}$ and the range (co-domain) $B = \{1,2\}$. Table 9.4 displays the $\#(B)^{\#A} = 2^3 = 8$ different ways of assigning images to the domain members and each of them—a column of the table—defines a different function on A to B. The fact should not be overlooked that there are only eight different functions on A to B and they are all displayed in the table.

TABLE 9.4

ALL FUNCTIONS ON $\{a,b,c\}$ to $\{1,2\}$

a	1	2	1	2	1	2	1	2
b	1	1	2	2	1	1	2	2
c	1	1	1	1	2	2	2	2

The number $\#(B)^{\#A}$ of distinct functions on A to B gives rise to the denotation

$$B^A = \text{the set of all functions } f : A \rightarrow B$$

even when B or A or both are not finite.

9.6 FUNCTIONS OF MORE THAN ONE
VARIABLE

This notation, and the number of distinct functions, are still the same when the domain is itself a Cartesian product; that is, members of the domain are themselves ordered pairs (or n-tuples, in higher dimensions) and the functions we are concerned with are **functions of two** (or more) **variables.**

As an example, consider the function which assigns a price to a particular quantity of an item of grocery. Thus, brand X of canned dog food may cost $2.00 for five cans but 43¢ a can. If the function name is p, these assignments may be written as $p(X,5) = 2.00$ and $p(X,1) = 0.43$. (Note that one set of parentheses vanishes in this notation, since the arguments are themselves ordered pairs, $(X,5)$ and $(X,1)$.)

Where $\mathbb{P}$ is the set of positive integers, as it has been used above, and F is the set of positive fractions, another example of a function of two variables is the function $r: \mathbb{P} \times \mathbb{P} \to F$ defined by

$$r(a,b) = \frac{a}{b}, \; \forall\, a,b \in \mathbb{P}.$$

It assigns the fraction as the value of the pair of integers (which are the numerator and denominator of the fraction). The domain of r is the Cartesian product $\mathbb{P} \times \mathbb{P}$, i.e. the set of all *ordered pairs* of positive integers and, strictly speaking, a member of the function r is of the form $((a,b),c)$, where $c = a/b$.

As a third example, we introduce a **binary operation** on a set S, defined as a function $f: S \times S \to S$, whose arguments are pairs of members of S and their images are members of S. A particular instance is the union of subsets of a set A:

$$f: \mathcal{P}(A) \times \mathcal{P}(A) \to \mathcal{P}(A)$$

defined by $f(B,C) = B \cup C, \; \forall\, B,C \subseteq A$.

In fact, all the familiar binary operations may be expressed in this fashion, and their common symbols may serve as the function names. For example, addition of real numbers may be depicted as

$$+: \mathbb{R} \times \mathbb{R} \to \mathbb{R}, \text{ defined by } +(a,b) = a + b,$$

$$\forall\, a,b \in \mathbb{R}.$$

When the domain is a Cartesian product of two different sets, say $f: A \times B \to C$, and all sets are finite, then $A \times B$ has $(\#A) \cdot (\#B)$ members and thus there are $(\#C)^{(\#A)(\#B)}$ different functions on $A \times B$ to C.

(We lose no generality when we consider functions of the form $f: D \to C$, since the domain D may well be a Cartesian product, without it being emphasized.)

9.7 MONIC (ONE-TO-ONE) FUNCTIONS

In general, a function is permitted to have the same value at different arguments. A function restricted so that distinct arguments must take distinct values has some special and useful properties.

9.5 Definition: A function $f: D \to C$ is **monic** (also **one-to-one**, an **injection**) iff

$$\forall\ x,y \in D,\ x \neq y \Rightarrow f(x) \neq f(y).$$

A function which is both monic and epic, or one-to-one and onto, is called a **bijection.** However, in this book the terms "monic" and "epic" are used.

One important property of monic functions is that their inverses are also functions. That is, the inverse relation (see Definition 7.3) also has the property that distinct first coordinates (which used to be second) have distinct second coordinates (which used to be first).

9.6 Theorem: Let $f: D \to R$ be a function and let $f(D) = R$ (i.e. the range of f is identical to its co-domain). Then f is a monic function iff $f^{-1}: R \to D$ is itself a function.

To elaborate a bit further, given a diagram illustrating a function f on a (geometric) coordinate system, we can obtain an illustration of the inverse f^{-1} by "flipping" the diagram about the 45° line, as is shown in Figure 9.7. Since the function f of part (i) is not monic, there is at least one case (an infinite

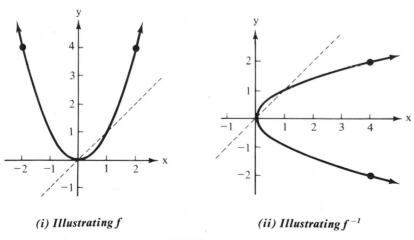

(i) Illustrating f   **(ii) Illustrating f $^{-1}$**

FIGURE 9.7

number of them, actually) of two distinct ordered pairs with the same second coordinate; for example, $(-2,4)$ and $(2,4)$. Then, in f^{-1} we have both pairs $(4,-2)$ and $(4,2)$, violating the restriction on a function.

9.8 THE SET FUNCTION INDUCED BY A RELATION

Even when a function is not monic, it is still possible to deal with its inverse as a function of sorts, provided we regard the inverse as a **set function**. This can be accomplished for any function, or relation in general, by a device that may be more easily grasped if illustrated before it is defined. We shall denote by $\mathbb{P}$ the set of positive integers.

Let $f: \mathbb{P} \to \mathbb{P}$ be defined by $f(p) = p(\text{mod } 3)$, $\forall p \in \mathbb{P}$; that is $f(p)$ is the remainder after dividing p by 3. Then, the domain of f is $\mathbb{P}$ and its range is $\{0,1,2\}$. We may now take any *subset* of $\mathbb{P}$, say $T = \{1,4,5\}$, collect the images of its members $(f(1) = 1, f(4) = 1, f(5) = 2)$ into a set $\{1,2\}$ and call it the f-**image of** T, written $f(T)$. Now, f was defined as a *point function* (the familiar function with individual members of the domain assigned members of the range to form ordered pairs), but now it is defined on a *subset of the domain* and its value is a *subset of the range*.

In fact, f may be defined in the same manner for *any* subset of the domain $\mathbb{P}$. For example, $f(\mathbb{P}) = \{0,1,2\}$, $f(\{2,5,8\}) = \{2\}, f(\{3x + 1: x \in \mathbb{P}\}) = \{1\}$, etc. We can mimic the original definition of f by its images on singleton sets: $f(\{1\}) = \{1\}, f(\{4\}) = \{1\}, f(\{15\}) = \{0\}$. The only statement still needed is that the image of the empty set is always empty, i.e. $f(\varnothing) = \varnothing$.

The illustration above could have taken place with a relation R that is not a function: define the R-image of a subset A of the domain of R as the set of second coordinates of all ordered pairs with first coordinates from A. In fact, the set function thus defined may use the same name as the relation, because the context prevents confusion.

> **9.7 Definition.**   Let R be a relation with domain D and range C. Then, the **set function** $R: \mathcal{P}(D) \to \mathcal{P}(C)$, **induced by** the relation R, is defined as follows: Let $A \subseteq D$, i.e. $A \in \mathcal{P}(D)$. Then $R(A) = \{y \in C: (x,y) \in R \text{ for some } x \in A\}$. The set $R(A)$ is called the R-**image of** A.

Thus the set function R is defined for each subset of D, including the empty set $\emptyset$. Even when the relation R is not a function, the *set function* R is one, since every subset A of D (member of the new domain $\mathcal{P}(D)$) has a *unique* image which is the set of images of A. When we start out with a point function f, then we have $f(A) = \{y: \text{for some } x \in A, y = f(x)\} = \{f(a): a \in A\}$. The set $f(A)$ is called the image of A under f, or the f-image of A.

9.8 Theorem: Let f be a set function and let A and B be members (sets) of its domain. Then,

(i) $f(A \cup B) = f(A) \cup f(B)$;

(ii) $f(A \cap B) \subseteq f(A) \cap f(B)$.

It should be noted in Theorem 9.8 (ii), that it is not true in general that $f(A \cap B) = f(A) \cap f(B)$, as is shown in Figure 9.8, where disjoint sets have intersecting images. The circles labeled $f(A)$ and $f(B)$ represent the f-images of the sets represented by the circles labeled A and B respectively, and both $a \in A$ and $b \in B$ are mapped to $c \in f(A) \cap f(B)$; i.e., $f(a) = f(b) = c$.

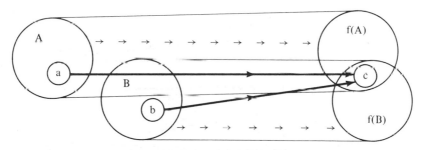

Disjoint sets may have intersecting images

FIGURE 9.8

9.9 THE INVERSE OF A FUNCTION

Where $f: D \to C$ is a function with $f(D) = C$, f^{-1} is not a function if f is not monic, but as a *set function*, $f^{-1}: \mathcal{P}(C) \to \mathcal{P}(D)$ is indeed a function. For any $A \subseteq C$, $f^{-1}(A)$ is called the **inverse (inverse image)** of A under f and is defined by

$$f^{-1}(A) = \{x \in D: f(x) \in A\},$$

that is, the collection of all arguments in D which f maps into A.

When the inverse image of a singleton is taken, the set-designating braces are often left out:

$$f^{-1}(y) = \{x \in D: f(x) = y\},$$

where y is any member of the range C. If the co-domain is larger than the range and y is in the former but not in the latter, then

$$f^{-1}(y) = \varnothing,$$

but f^{-1} is still defined at y. The same is true, of course for $f^{-1}(A)$ when $A \cap$ range $f = \varnothing$.

It is interesting to observe how the inverse of a function behaves on the original domain of the function.

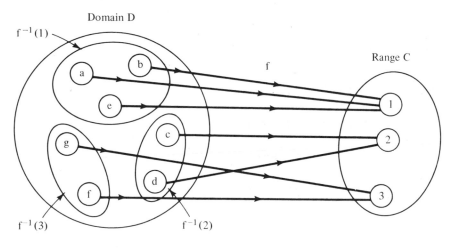

The inverse images of singletons are disjoint

FIGURE 9.9

Since each member of the domain is assigned a unique member of the range, the inverse image of a singleton set from the range must include all, and only, those domain members which were mapped to it. For example, in Figure 9.9, *a,b*, and *e* are all the elements mapped to 1 and therefore $f^{-1}(\{1\}) = \{a,b,e\}$. Moreover, none of these three domain members can be in $f^{-1}(\{2\})$ or $f^{-1}(\{3\})$ because they were not mapped to either 2 or 3. We thus have the following:

9.9Theorem: If f is a function, then $f^{-1} \circ f$ is an equivalence relation on the domain of f.

A brief glimpse shows the reason:

$$(x,y) \in f^{-1} \circ f \text{ iff } f(x) = f(y).$$

It is of interest to note that, when f is a function, the composition $f \circ f^{-1}$ is a function. A glance at Figure 9.9 shows that it is the identity on the range of f.

The result in Theorem 9.9 explains why in Theorem 9.8(ii) we could do no better than the inclusion $f(A \cap B) \subseteq f(A) \cap f(B)$, while with the inverse f^{-1} we have equality. The reason is that f^{-1} causes the separation of the domain into (*disjoint*) equivalence classes, eliminating the interference shown in Figure 9.8.

A similar situation holds for the relative complement, or set difference. The best that can be done in the general case is set inclusion.

9.10 Theorem: Let $f: D \to C$ be a function and let $A,B \subseteq D$. Then,

$$f(A) - f(B) \subseteq f(A - B).$$

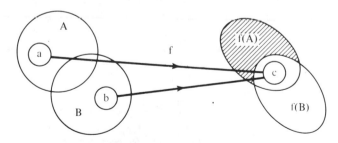

The reason equality does not hold in Theorem 9.10

FIGURE 9.10

The reason equality does not always hold is that an element $a \in A - B$ may be mapped by f to $c = f(b)$, for some $b \in B$. Then c will be removed from $f(A)$ when $f(B)$ is removed; yet $a \in B$, and thus $c = f(a) \in f(A - B)$. However, the disjointness of the equivalence classes induced by f^{-1} yields equality when f is replaced by f^{-1}. The following theorem states that the inverse of a function preserves relative complements, unions, and intersections.

9.11 Theorem: Let f be a function and let A and B be subsets of its range. Then,

(i) $f^{-1}(A - B) = f^{-1}(A) - f^{-1}(B)$.

(ii) $f^{-1}(A \cup B) = f^{-1}(A) \cup f^{-1}(B)$.

(iii) $f^{-1}(A \cap B) = f^{-1}(A) \cap f^{-1}(B)$.

Parts (ii) and (iii) of the theorem are easily extendable to the union and intersection of an arbitrary family of sets.

9.10 RESTRICTIONS AND EXTENSIONS OF FUNCTIONS

Restrictions and extensions were defined in Section 7.20 for relations. Here we present the two concepts for the special case of functions.

We recall that a function is defined by specifying its domain and the value of the function at each member of the domain. Thus, the two "squaring functions," $f \colon \mathbb{R} \to \mathbb{R}$ defined by $f(x) = x^2$ for each $x \in \mathbb{R}$, and $g \colon \mathbb{P} \to \mathbb{P}$ defined by $g(x) = x^2$ for each $x \in \mathbb{P}$, are not the same function, since they have *different domains*. Yet, whenever g is defined, so is f, because $\mathbb{P} \subset \mathbb{R}$; moreover, the values of f and g are identical at each member of $\mathbb{P}$. In other words, the domain of g is part of the domain of f, and on that part the two functions "behave" identically. We then say that g is a **restriction** of f to $\mathbb{P}$; we also say that f is an **extension** of g to $\mathbb{R}$.

9.12 Definition: Let $f \colon A \to B$ be a function and let $C \subseteq A$. Then $f \cap (C \times B)$ is a function called the **restriction** of f to C, denoted $f|_C$ (or $f | C$). Its domain is C, and $(f|_C)(x) = f(x)$, for every $x \in C$.

A function g is the **restriction** of f to D iff $D \subseteq A$, D is the domain of g, and $g(x) = f(x)$, $\forall x \in D$. Thus g is a restriction of f iff $g \subseteq f$. The function f is an **extension** of g iff g is the restriction of f to some subset of the domain of f.

10

Special Functions

There are several functions and families of functions which play important roles in mathematics and computer science. We discuss a few of the more prominent ones in this section.

10.1 THE IDENTITY FUNCTION

The domain and range of the identity function are the same set. The value which the function assigns to each member of the domain is that very member. The notation used for the identity function varies, but it usually includes a mention of its domain. The identity on A is commonly denoted 1_A, i_A, I_A, e_A, and id_A. We adopt the latter.

id_A

10.1 Definition: Let A be a set. The **identity function** on A is $id_A: A \rightarrow A$ defined by

$$id_A(x) = x, \text{ for every } x \in A.$$

As was seen earlier, for any function $f: A \rightarrow B$, with $f(A) = B$, the composition $f \circ f^{-1}: B \rightarrow B$ is the identity on B; i.e., $f \circ f^{-1} = id_B$.

If $f: A \rightarrow B$ is a monic and epic function, then $f \circ f^{-1} = id_B$ and $f^{-1} \circ f = id_A$ (with appropriate allowances made for the conversion between point functions and set functions).

10.2 THE NATURAL FUNCTION

When a set A is partitioned into equivalence classes, it is "natural" to associate an element of A with the equivalence class to which it belongs. This amounts to assigning the equivalence class as a value to the element. The function which accomplishes this has A as its domain and the partition as its range. This function is called the **natural (selection, membership, partitioning) function** and it often is accompanied by the special notation $[x]$, denoting the equivalence class to which x belongs.

> **10.2 Definition:**   Let A be a (nonempty) set, let $\pi = \{\pi_i: i \in I\}$ be a partition of A for some nonempty indexing set I, and for any $x \in A$ let $[x]$ denote that member π_i of π to which x belongs.
> The **natural function** induced by π on A is $f_\pi: A \to \pi$ defined by
>
> $$f_\pi(x) = [x], \text{ for each } x \in A.$$

The natural function is a convenient tool in dealing with partitions and equivalence classes. For example, $f_\pi(x) = f_\pi(y)$ expresses in functional notation the fact that x and y are in the same equivalence class, that is $[x] = [y]$. This notation may be further used for chaining, or linking, partitions and using functional composition for successive refinements.

10.3 THE CHARACTERISTIC FUNCTION

The **characteristic function** makes it possible to represent numerically each subset of a universal set. For each such subset, there is a unique characteristic function, and the domain of all these characteristic functions is conveniently the same—the universal set.

χ_A

Let us illustrate with the universal set $U = \{a,b,c,d,e\}$. The characteristic function of a subset A of U, denoted by χ_A, assigns to each element x of the *universal set* U the value 1 if $x \in A$ and the value 0 otherwise. For example, let $A = \{a,b,c\}$ and let $B = \{b,d\}$, then χ_A is defined by

x	a	b	c	d	e
$\chi_A(x)$	1	1	1	0	0

where χ_B is defined by

x	a	b	c	d	e
$\chi_B(x)$	0	1	0	1	0

10.3 Definition: Let U be a nonempty set and let $A \subseteq U$. Then the **characteristic function** of A is $\chi_A: U \rightarrow \{0,1\}$ defined for every $x \in U$ by,

$$\chi_A(x) = \begin{cases} 1, \text{ if } x \in A, \\ 0, \text{ if } x \notin A. \end{cases}$$

It is clear from the definition that $\chi_\phi(x) = 0$, for every $x \in U$, and that $\chi_U(x) = 1$, for every $x \in U$. Another feature of the characteristic function is that there is a 1–1 correspondence between the subsets of U and the characteristic functions over U; that is, every subset of U has its own *unique* characteristic function and, once the images of the members of U have been fixed each at 0 or 1, the resulting function describes a unique subset of U—the one with those elements whose image is 1.

One of the most convenient feature of the definition is that all the characteristic functions of subsets of U have the same domain U and thus they may be treated in a uniform manner. Also, the fact that the value 0 signals the absence of the argument from the subset and the value 1 its presence, allows multiplication and addition to simplify the manipulations of sets considerably.

For example, consider the function f which assigns to a positive integer the remainder after dividing by 3. Some samples of its values are: $f(1) = 1, f(2) = 2, f(3) = 0, f(4) = 1, f(5) = 2$, etc. We can define the function f with the aid of the characteristic functions on subsets of $\mathbb{P}$ in the following manner. Let

$$A_0 = \{3,6,9,12, \ldots\} = \{p \in \mathbb{P}: p = 0(\text{mod } 3)\},$$
$$A_1 = \{1,4,7,10, \ldots\} = \{p \in \mathbb{P}: p = 1(\text{mod } 3)\}, \text{ and}$$
$$A_2 = \{2,5,8,11, \ldots\} = \{p \in \mathbb{P}: p = 2(\text{mod } 3)\}.$$

Thus A_i is the set of all positive integers with remainder i after division by 3, $\forall i \in \{0,1,2\}$. Also, these three sets exhaust all positive integers, i.e. $\mathbb{P} = \bigcup_{i=0}^{2} A_i = A_0 \cup A_1 \cup A_2$.

Now

$$\chi_{A_0}(p) = \begin{cases} 1, \text{ if } p \text{ is divisible by 3,} \\ 0, \text{ otherwise.} \end{cases}$$

Also, $$\chi_{A_1}(p) = \begin{cases} 1, \text{ if } p = 1(\text{mod } 3), \\ 0, \text{ otherwise.} \end{cases}$$

And $$\chi_{A_2}(p) = \begin{cases} 1, \text{ if } p = 2(\text{mod } 2), \\ 0, \text{ otherwise.} \end{cases}$$

For each positive integer p, exactly one of the three functions will have the value 1 and the other two will have the value 0. For instance $\chi_{A_0}(4) = 0$, $\chi_{A_1}(4) = 1$, and $\chi_{A_2}(4) = 0$.

We are now ready to define the function $f: \mathbb{P} \rightarrow \{0,1,2\}$. For each positive integer p,

$$f(p) = \sum_{i=0}^{2} i\chi_{A_i}(p) = 0\chi_{A_0}(p) + 1\chi_{A_1}(p) + 2\chi_{A_2}(p).$$

Thus, $f(7) = 0\chi_{A_0}(7) + 1\chi_{A_1}(7) + 2\chi_{A_2}(7) = 0 \cdot 0 + 1 \cdot 1 + 2 \cdot 0 = 1$; and $f(8) = 0 \cdot 0 + 1 \cdot 0 + 2 \cdot 1 = 2$, since $\chi_{A_0}(8) = \chi_{A_1}(8) = 0$, and $\chi_{A_2}(8) = 1$.

The preceding is a rather simple example for the use of the characteristic functions. They are used to great benefit in proofs in areas such as transfinite cardinality, probability theory and others. Here we shall conclude by illustrating the use of the characteristic function to translate operations on subsets into arithmetic.

The characteristic functions of a set and its complement must have opposite values for each argument, since $x \in A$ iff $x \notin A'$ and therefore $\chi_A(x) = 0$ iff $\chi_{A'} = 1$ and $\chi_A(x) = 1$ iff $\chi_{A'} = 0$. The result is the identity

$$\chi_{A'}(x) = 1 - \chi_A(x), \forall x \in U.$$

We now translate the union and intersection of two sets with the use of characteristic functions. The Venn diagram in Figure 10.1 will be of help. In it, each of the circles representing the two sets A and B has been shaded in a different direction so that $A \cup B$ is represented by the total shaded area and $A \cap B$ is represented by the cross-hatched area.

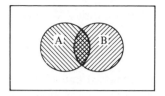

Venn diagram

FIGURE 10.1

Now,

$$\chi_A(x) = \begin{cases} 1, \text{if } x \in A, \\ 0, \text{if } x \notin A; \end{cases}$$

and

$$\chi_B(x) = \begin{cases} 1, \text{if } x \in B, \\ 0, \text{if } x \notin B. \end{cases}$$

For an element $x \in A \cap B$, i.e. in the cross-hatched area, we have both $\chi_A(x) = 1$ and $\chi_B(x) = 1$. We note therefore that the product of the two functions is 1 only for members of the intersection; i.e.

$$\chi_A(x)\chi_B(x) = \begin{cases} 1, \text{if } x \in A \cap B; \\ 0, \text{otherwise.} \end{cases}$$

But that yields the following representation for the characteristic function of the intersection.

$$\chi_{A \cap B}(x) = \chi_A(x) \cdot \chi_B(x), \quad \forall x \in U.$$

On the other hand, if we added $\chi_A(x)$ and $\chi_B(x)$, as long as $x \in A - B$ or $x \in B - A$, the result is 1, and when $x \notin A \cup B$ the result is 0, all of which are desired results. But when $x \in A \cap B$ it is counted twice and the sum is 2. Hence, we subtract $\chi_A(x)\chi_B(x)$ from the sum, which subtracts 0 when there is no duplication, and 1 when reduction should occur. Consequently,

$$\chi_{A \cup B}(x) = \chi_A(x) + \chi_B(x) - \chi_A(x)\chi_B(x), \forall x \in U.$$

The identities obtained above may be used to prove set-algebraic identities. We shall use the fact that

$$\chi_A(x) \cdot \chi_A(x) = \chi_A(x),$$

for all subsets A and all $x \in U$ (because $0 \cdot 0 = 0$ and $1 \cdot 1 = 1$). Then the absorption law $A \cap (A \cup B) = A$ may be verified by,

$$
\begin{aligned}
\chi_{A \cap (A \cup B)}(x) &= \chi_A(x)\chi_{A \cup B}(x) \\
&= \chi_A(x)(\chi_A(x) + \chi_B(x) - \chi_A(x)\chi_B(x)) \\
&= \chi_A(x)\chi_A(x) + \chi_A(x)\chi_B(x) - \chi_A(x)\chi_A(x)\chi_B(x) \\
&= \chi_A(x) + \chi_A(x)\chi_B(x) - \chi_A(x)\chi_B(x) \\
&= \chi_A(x).
\end{aligned}
$$

The reader may find it of interest to try this device on another of the laws of operating with sets, which appear in Table 4.1.

It is also of interest to note of the inverse of the characteristic function that

$$\chi_A^{-1}(1) = A \text{ and } \chi_A^{-1}(0) = A'.$$

10.4 PERMUTATIONS

[Permutations in the context of combinatorial analysis are discussed in Section 5.11. Here, permutations are viewed as functions on a set of "marks."]

The intuitive notion of a permutation on a set is that of rearranging its elements. An original order of the elements is assumed and then the permutation changes the order, or leaves it the same. Formally, a permutation is a *monic and epic function on a set*. Commonly, and exclusively in this book, the set is taken to be finite. The permutation is listed in any one of several ways, but it must define the image of each member of the set.

For example, permutations on the set $\{a,b,c\}$ may be listed in a table, where each column defines a permutation (function) by listing the images of the elements from the label column, as is done in Table 10.1. The first column in part (i) defines the **identity permutation,** which does not change the order. The

TABLE 10.1

PERMUTATIONS ON {a,b,c}

	1	2	3	4	5	6			1	2	3	4	5	6
a	a	a	b	b	c	c		b	b	c	a	c	a	b
b	b	c	a	c	a	b		c	c	b	c	a	b	a
c	c	b	c	a	b	a		a	a	a	b	b	c	c

(i)   (ii)

remaining five columns define the remaining permutation on {a,b,c}. The columns were numbered for convenience only.

Part (ii) of the table defines exactly the same permutations in the same order, as a moment's inspection will show. This points out the need for caution in reading and writing permutations.

For each of the permutations there is an **inverse** in the table, since a permutation is a monic and epic function. The identity permutation is its own inverse, as are the ones in columns 2, 3, and 6. The permutations in columns 4 and 5 are each other's inverses: applying column 4 first, followed by column 5, a is changed to b then b to a, etc.

Another common notation for permutations is the one listing the elements in the original order in the first row and their images in the second. Thus, the permutations of Table 10.1 will appear as

$$\begin{pmatrix} a & b & c \\ a & b & c \end{pmatrix}, \begin{pmatrix} a & b & c \\ a & c & b \end{pmatrix}, \begin{pmatrix} a & b & c \\ b & a & c \end{pmatrix},$$

$$\begin{pmatrix} a & b & c \\ b & c & a \end{pmatrix}, \begin{pmatrix} a & b & c \\ c & a & b \end{pmatrix}, \begin{pmatrix} a & b & c \\ c & b & a \end{pmatrix}.$$

What is probably the most convenient form of displaying a permutation is one which encloses in parentheses the list of elements, each of which has its right-hand neighbor as its image, and the image of the last in the parentheses is the first in the parentheses. Table 10.2 displays the correspondence between the two forms.

To achieve the second form from the first, *open parentheses and write the first unused letter from the top row; then follow it with its image in the bottom row, unless that image

TABLE 10.2

THE PERMUTATIONS ON $\{a,b,c\}$

$$\begin{pmatrix} a & b & c \\ a & b & c \end{pmatrix} = (a)\,(b)\,(c), \qquad \begin{pmatrix} a & b & c \\ b & c & a \end{pmatrix} = (abc),$$

$$\begin{pmatrix} a & b & c \\ a & c & b \end{pmatrix} = (a)\,(bc), \qquad \begin{pmatrix} a & b & c \\ c & a & b \end{pmatrix} = (acb),$$

$$\begin{pmatrix} a & b & c \\ b & a & c \end{pmatrix} = (ab)(c), \qquad \begin{pmatrix} a & b & c \\ c & b & a \end{pmatrix} = (ac)\,(b).$$

already appears in the parentheses and in that case close the parentheses. If not all letters are used, return to the * above; otherwise, stop.

There is considerable volume of algebraic study associated with permutations, since they form a certain kind of group with rather strong properties. (The subject of Section 12.8 is permutation groups.) Here we only remark on the number of distinct permutations of n letters. The argument is similar to that in Section 9.5 concerning the number of distinct functions on A to B. The difference here is that once a letter has been used as an image, it cannot be used again in the same permutation.

There are then n alternatives for the first image, but once it has been chosen, no matter what it is, there are only $n-1$ alternatives available. Each position has a number of alternatives smaller by one, until the last image must be the only remaining letter. Consequently there are

$$n \cdot (n-1) \cdot (n-2) \cdot \ldots 2 \cdot 1 = n!$$

(read "n factorial") distinct permutations on n letters. This should serve as a caution, as $n!$ grows very large very quickly:

$$2! = 2, \ 3! = 6, \ 4! = 24, \ 5! = 120, \text{ and } 10! = 3,628,800.$$

10.5 FUNCTIONS WHICH PRESERVE PROPERTIES

Often the domain of a function is a set on which some structure or property has been established, such as an ordering (partial or linear), a binary operation, and the like. The question can then be asked whether the function preserves the

TABLE 10.3

DEFINING $ ON {0,1,2}

$	0	1	2
0	0	1	2
1	1	2	0
2	2	0	1

property as it maps the domain elements into those of the range. For instance, suppose $f\colon A \to B$ is a function, suppose $(A,\leq)$ is a poset, and suppose $(B,\lessdot)$ is a poset. Let $a \leq b$ for some $a,b \in A$. If the function f *preserves* the partial ordering, then $f(a) \lessdot f(b)$ must be true, independent of the choice of a and b.

In the following example the function preserves the binary operation. Let $+\colon \mathbb{P} \times \mathbb{P} \to \mathbb{P}$ be the common addition of positive integers. Let \$ be the binary operation on {0,1,2} which is addition modulo 3, as displayed in Table 10.3. Now define $f\colon \mathbb{P} \to \{0,1,2\}$ as the function which assigns to a positive integer p the remainder after dividing p by 3. (In the discussion of characteristic functions f was defined as $f(p) = 0\chi_{A_0}(p) + 1\chi_{A_1}(p) + 2\chi_{A_2}(p)$.) It then is the case that f preserves the addition on $\mathbb{P}$, in the following sense: For any two positive integers a and b, on the one hand add them in $\mathbb{P}$ and then find the f-image $f(a + b)$ in {0,1,2}, and on the other, map a and b first to obtain their images $f(a)$ and $f(b)$, and then perform the operation \$ on the images to obtain $f(a) \$ f(b)$. The two results will always be the same; that is,

$$f(a + b) = f(a) \$ f(b), \quad \forall a,b \in \mathbb{P}.$$

As an example, let $a = 16$ and $b = 31$. Then, $f(a + b) = f(47) = 2$, while $f(a) \$ f(b) = 1 \$ 1 = 2$. Whenever a function preserves the operation, or other properties, it is called a **homomorphism.** Such functions are more frequently encountered in the study of algebra and are further discussed in Chapter 13.

10.6 MONOTONIC FUNCTIONS

There are several types of order preservations, of which we briefly discuss only the preservation of linear ordering by

monotonic functions. For example, the function f: $\mathbb{R} \rightarrow \mathbb{R}$ such that $f(x) = 2x + 3$ has larger values for larger arguments, and is therefore **monotonic increasing;** while g: $\mathbb{R} \rightarrow \mathbb{R}$ such that $g(x) = -x + 1$ has smaller values for bigger arguments, and is therefore **monotonic decreasing.**

As another example, consider the function h: $\mathbb{R} \rightarrow \mathbb{R}$ such that

$$h(x) = \begin{cases} x + 1, \text{ if } x < 0, \\ 1, \text{ if } x \geq 0. \end{cases}$$

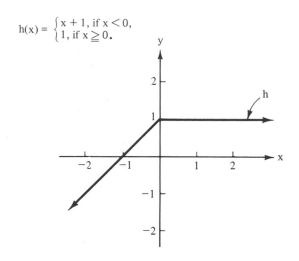

Illustrating h

FIGURE 10.2

As may be seen in Figure 10.2, for *larger* arguments h never takes smaller values; the values are either the same, if $x \geq 0$, or larger, if $x < 0$. Such a function is called **monotonic nondecreasing.** The analogy to a **monotonic nonincreasing** function should be obvious at this point.

In the illustrations just given, the linear orderings in the domain and the range were the same. The definition, below, considers the more general case.

10.4 Definition: Let f: $A \rightarrow B$ be a function, let $\leq$ be a reflexive linear order on A, let $\preceq$ be a reflexive linear order on B, and let $a,b \in A$. Then, f is said to be

(i) **monotonic nondecreasing** iff $a < b$ implies $f(a) \preceq f(b)$;

(ii) **monotonic nonincreasing** iff $a < b$ implies $f(b) \preceq f(a)$;

(iii) **monotonic increasing** iff $a < b$ implies $f(a) \prec f(b)$;

(iv) **monotonic decreasing** iff $a < b$ implies $f(b) \prec f(a)$.

10.7 BOOLEAN FUNCTIONS: EXPRESSIONS FROM VALUES

It is often the case in applications of Boolean algebra that tables of values for Boolean functions of several variables are obtained. In the case of logical design of circuits, these tables are often translated into "and-or-not" Boolean expressions which are, in turn, translated into the design of circuits.

In this section, we present two techniques for obtaining the desired Boolean expressions from the function values, with as little background information as we can manage.

Let us start with an example, trace back from it as far as necessary, and then move forward with it to introduce the desired techniques. We are given a table which lists the values of five functions f, g, h, j and k, each of them a function of three independent variables x, y and z. The functions are called **Boolean functions** because they take values in $\{0,1\}$ and each of their variables takes values in $\{0,1\}$. (This is not quite as general a definition as it could be, but it will do for this treatment.) The dashes in place of values in Table 10.4 are explained presently.

TABLE 10.4

FIVE BOOLEAN FUNCTIONS

x	y	z	f	g	h	j	k
0	0	0	0	—	0	—	0
0	0	1	0	—	0	1	0
0	1	0	0	1	0	—	0
0	1	1	0	1	0	1	0
1	0	0	0	—	1	0	0
1	0	1	1	0	—	0	0
1	1	0	—	0	0	—	1
1	1	1	—	0	0	1	0

Let us denote $\{0,1\}$ by B. Then, each of the five functions has $B \times B \times B$ as domain and B as range, e.g. $f: B \times B \times B \to B$; the value $f(x,y,z)$ of f at a combination (x,y,z) of values (e.g. $f(1,1,0)$) is listed in the table as 0 or 1, or as a dash. A *dash is used when it does not matter* whether the value is 0 or 1, when we "don't care" which it is; the technical term for the dash as a value is **don't care**.

For example, $f(0,0,1) = 0$, $f(1,0,1) = 1$, and $f(1,1,1) = —$, in Table 10.4 are values which f takes at *some* of the combinations of values for the independent variables. Now, since each of the variables $x, y,$ and z assumes as values either 0 or 1, there are $2^3 = 8$ combinations of values for them, the ones listed on the left of the vertical line in the table. *Our task is to produce a Boolean expression in terms of $x, y,$ and z which* takes the same value as f at every element of the domain (i.e. combination of 0's and 1's for x, y, and z). Such Boolean expressions will also have values (0 or 1) when the value of f is "don't care," but that is both acceptable and necessary.

By a **Boolean expression** here is meant an expression using

x, y, z as variables,

0,1 as constants,

$\vee, \wedge$ as binary operations,

$^-$ (over-bar) as a unary operation,

and grouping symbols.

The rules are some of those of Table 4.1 with the following correspondence:

Set notation	$\phi,\ U,\ \cup,\ \cap,\ '$
Boolean notation	$0,\ 1,\ \vee,\ \wedge,\ ^-$

(There is a corresponding concept $\leq$ in Boolean expressions to the subset $\subseteq$ relation in manipulation of sets, but we shall not use it here.)

The names for the operations vary from user to user. The operation $\vee$ (as in $x \vee y$) is called **or, inclusive or, join, disjunction,** and **alternation.** The operation $\wedge$ (as in $x \wedge y$) is called **and, meet, conjunction, product,** and **multiplication.** The unary operation $^-$ (as in $\bar{x}$) is called **not, negation, complement,** and **inversion.**

The basic laws of Boolean algebra are summarized in Table 10.5.

The tables for the three Boolean operations are easily derivable from the rules in Table 10.5. They are given in Figure 10.3 for the reader's convenience.

An example of a Boolean expression representing f from Table 10.4 is

$$f = x \wedge z.$$

TABLE 10.5

SUMMARY OF BOOLEAN LAWS

$$a \vee b = b \vee a \qquad a \wedge b = b \wedge a$$

$$(a \vee b) \vee c = a \vee (b \vee c) \qquad (a \wedge b) \wedge c = a \wedge (b \wedge c)$$

$$a \wedge (b \vee c) = (a \wedge b) \vee (a \wedge c) \qquad a \vee (b \wedge c) = (a \vee b) \wedge (a \vee c)$$

$$a \vee a = a \qquad a \wedge a = a$$

$$a \vee 0 = a \qquad a \wedge 1 = a$$

$$a \vee 1 = 1 \qquad a \wedge 0 = 0$$

$$a \vee \bar{a} = 1 \qquad a \wedge \bar{a} = 0$$

$$\overline{a \vee b} = \bar{a} \wedge \bar{b} \qquad \overline{a \wedge b} = \bar{a} \vee \bar{b}$$

$$\overline{(\bar{a})} = a$$

$$\bar{0} = 1 \qquad \bar{1} = 0$$

$\vee$	0	1
0	0	1
1	1	1

$\wedge$	0	1
0	0	0
1	0	1

x	$\bar{x}$
0	1
1	0

Definition of Boolean operations

FIGURE 10.3

With the use of the tables in Figure 10.3, the reader may verify this equality for each combination of values of the variables. Here we shall demonstrate only two: When $x = 0$ and $z = 1$, the value listed for f in Table 10.3 is 0; also $0 \wedge 1 = 0$. When $x = 1 = z$, we have $f = 1$ in Table 10.3 and $1 \wedge 1 = 1$.

We shall use the common convention permitting us to omit the conjunction $\wedge$ and to replace it by a dot, or to omit it altogether. Thus, we may write

$$h = x \wedge \bar{y} = x \cdot \bar{y} = x\bar{y}.$$

(The reader could verify this expression for h of Table 10.3.)

The reason it is helpful to translate the table of values of a function into a Boolean expression (there are many such expressions for each function) is the fact that the three operations, $\vee, \wedge, ^-$, may be imitated by electronic circuit elements. Consequently, a Boolean function may be **realized** as a circuit, and that is its use in several applications to computer science.

The imitation is done with pieces of equipment called an OR-element, an AND-element, and a NOT-element. Regarding the Boolean value 0 as "no current in the wire" or "low voltage", and the value 1 as "there is current in the wire" or "high voltage", the effects of these three elements are displayed in Figure 10.4.

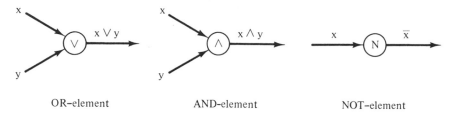

OR–element   AND–element   NOT–element

Logic elements

FIGURE 10.4

Now that the context has been explained, it remains to accomplish the translation of a list of 0's and/or 1's into a Boolean expression.

The simplest, but not most efficient, method uses **complete products.** In our case, a complete product is an AND-product of three letters,

the first is either x or $\overline{x}$,

the second is either y or $\overline{y}$,

and the third is either z or $\overline{z}$.

A list of all the complete products in our case follows, with $\wedge$'s omitted:

$$\overline{x}\,\overline{y}\,\overline{z}, \quad \overline{x}\,\overline{y}\,z, \quad \overline{x}\,y\,\overline{z}, \quad \overline{x}\,y\,z,$$

$$x\,\overline{y}\,\overline{z}, \quad x\,\overline{y}\,z, \quad x\,y\,\overline{z}, \quad x\,y\,z.$$

A complete product then has each of the variables as a factor exactly once, each with or without the complement bar.

Since a complete product is an AND-product ($\wedge$-product), it has the value 1 only when the value of each of its factors is 1. Thus, for example, $x\,\bar{y}\,\bar{z}$ has the value 1 only when $x = 1, \bar{y} = 1$, and $\bar{z} = 1$. But that happens exactly when $x = 1$, $y = 0$, and $z = 0$. For all other seven combinations of values for x, y, and z, (the remaining seven rows of Table 10.4), $x\,\bar{y}\,\bar{z}$ has the value 0. To reiterate, *every complete product has the value 1 for exactly one combination of values for its variables.*

Consequently, we can match to each 1-entry for a function in the table the unique complete product which has the value 1 for the same combination of values. For example, the function k of Table 10.4 has the value 1 only for $(x,y,z) = (1,1,0)$. The complete product that has the value 1 there is $x\,y\,\bar{z}$, and therefore we may write

$$k = x\,y\,\bar{z}.$$

For any other combination of values, say the third line of the table where $(x,y,z) = (0,1,0)$, the value of $x\,y\,\bar{z}$ is 0:

$$(x\,y\,\bar{z})(0,1,0) = 0 \cdot 1 \cdot \bar{0} = 0 \cdot 1 \cdot 1 = 0.$$

TABLE 10.6

COPY OF EARLIER TABLE WITH
MATCHING COMPLETE PRODUCTS

Matching complete product	x	y	z	f	g	h	j	k
$\bar{x}\,\bar{y}\,\bar{z}$	0	0	0	0	—	0	—	0
$\bar{x}\,\bar{y}\,z$	0	0	1	0	—	0	1	0
$\bar{x}\,y\,\bar{z}$	0	1	0	0	1	0	—	0
$\bar{x}\,y\,z$	0	1	1	0	1	0	1	0
$x\,\bar{y}\,\bar{z}$	1	0	0	0	—	1	0	0
$x\,\bar{y}\,z$	1	0	1	1	0	—	0	0
$x\,y\,\bar{z}$	1	1	0	0	0	—	1
$x\,y\,z$	1	1	1	0	0	1	0

Now, if the values listed for the function include more than one 1, as in g of the table, we may list for each value 1 the corresponding complete product (whose value is 1 for that same combination) and combine all such complete products in an **OR-sum.** For g, these complete products are $\bar{x}\,y\,\bar{z}$ (for the third row) and $\overline{x} \, y \, z$ (for the fourth row). Thus,

$$g = \overline{x} \, y \, \overline{z} \vee \overline{x} \, y \, z.$$

The reason this would work is the following: for every row of the table, other than the third and the fourth, both complete products have the value 0 and so we have $0 \vee 0 = 0$ for g. Since one or the other complete products has the value 1 for the third and fourth rows of the table, g has the value 1 there.

In the same manner we could obtain:

$$f = x \, \overline{y} \, z,$$

$$h = x \, \overline{y} \, \overline{z},$$

and

$$j = \overline{x} \, \overline{y} \, z \vee \overline{x} \, y \, z \vee x \, y \, z.$$

We already have:

$$g = \overline{x} \, y \, \overline{z} \vee \overline{x} \, y \, z$$

and

$$k = x \, y \, \overline{z}.$$

These five Boolean expressions form a *correct answer*— they accomplish the given task. However, these are often inefficient and will result in unnecessarily complicated circuits; it is desirable to **reduce** these expressions. By that is often meant, to reduce the number of **literals** (a variable or its complement) in the expression (although that is not always the most desirable outcome).

One reduction method is **algebraic simplification,** in which the rules of Table 10.5 are used. We leave it to the reader to trace the details, but the results for g and j may be as follows:

$$g = \overline{x} \, y \, \overline{z} \vee \overline{x} \, y \, z$$
$$= \overline{x} \, y (\overline{z} \vee z)$$
$$= \overline{x} \, y \cdot 1$$
$$= \overline{x} \, y.$$
$$j = \overline{x} \, \overline{y} \, z \vee \overline{x} \, y \, z \vee x \, y \, z$$
$$= (\overline{x} \, \overline{y} \vee \overline{x} \, y \vee x \, y) z$$
$$= (\overline{x} \, \overline{y} \vee (\overline{x} \, y \vee \overline{x} \, y) \vee x \, y) z$$

$$= ((\overline{x}\,\overline{y} \vee \overline{x}\,y) \vee (\overline{x}\,y \vee x\,y))z$$
$$= (\overline{x}(\overline{y} \vee y) \vee (\overline{x} \vee x)y)z$$
$$= (\overline{x}\cdot 1 \vee 1\cdot y)z$$
$$= (\overline{x} \vee y)z.$$

Admittedly, algebraic simplification is not always easy, nor does it furnish a blueprint that may be easily followed. There are several algorithmic methods which aid in simplification (often called minimization) of Boolean functions, but which are too involved to be included here. Three of them are, the **map** method (**Veitch-Karnaugh**), the **Quine-McCluskey** method, and the **consensus** method. (The interested reader should consult the reading list.)

Even without these methods, we may still improve on the results above by using the freedom afforded by the *don't care* values in the table. The main idea is that we may assign the value 1 instead of a don't care whenever that would simplify the expression. For example, the expression we have for f is $f = x\,\overline{y}\,z$. If we now assigned 1 as a value of f for the combination in the last row of the table, we would express f as

$$f = x\,\overline{y}\,z \vee x\,y\,z$$

and simplify it, to get

$$f = x(\overline{y}\,z \vee y\,z)$$
$$= x(\overline{y} \vee y)z$$
$$= x\cdot 1\cdot z$$
$$= x\,z.$$

Similarly, we obtain $g = \overline{x}$ by assigning 1 to g for the first two combinations ($g(0,0,0) = g(0,0,1) = 1$); $h = x\,\overline{y}$, by assigning 1 to h at the sixth combination ($h(1,0,1) = 1$); and $j = \overline{x} \vee y$, by assigning 1 to j at all three don't care positions ($j(0,0,0) = j(0,1,0) = j(1,1,0) = 1$).

Although it is somewhat of an art to become aware of such possibilities, one may use as a guide the following idea: when two value combinations differ in only one position, e.g. $(0,0,1)$ and $(1,0,1)$, the complete products whose values are 1 at these combinations differ in one variable only, and in a particularly helpful way: one letter appears complemented in one complete product and uncomplemented in the other. In our example, x appears complemented in $\bar{x}\,\bar{y}\,z$ and uncomplemented in $x\,\bar{y}\,z$, while the remaining variables are identical. But then, in the OR-sum of the two complete products x and $\bar{x}$ *will vanish*, leaving only one product of only two literals:

$$\bar{x}\,\bar{y}\,z \vee x\,\bar{y}\,z = (\bar{x} \vee x)\bar{y}\,z = 1 \cdot \bar{y}\,z = \bar{y}\,z.$$

This idea holds true for combining shorter products as well, and the reader should remain alert to it.

The last helpful hint is the fact that any product may be repeated in the OR-sum because of idempotency, i.e. $u \vee u = u$ for any u. (We have actually used this in the first simplification of j.)

Armed with these largely seat-of-the-pants tools, one may accomplish the desired reduction in applications mainly by inspection. We grant that inspection is not an algorithmic process, but then we can always fall back on the *OR-sum of all complete products* corresponding to 1-values of the function, which is a correct solution, *obtained algorithmically*; it is just not always pleasant.

The reader who desires a more systematic minimization technique has to be willing to spend the time and effort required by a method such as indicated above.

11

Simple Algebraic Structures—A Set and an Operation

11.1 BINARY OPERATIONS AND CLOSURE

Some algebraic systems are more complex, and have more components, than the ones considered here. For our purposes, it will suffice to consider a set and a binary operation defined on the set.

Some familiar examples are:

1. $(\mathbb{P},+)$, the set of positive integers with ordinary addition.
2. $(\mathbb{R},\cdot)$, the set of real numbers with ordinary multiplication.
3. $(\{1,2,\ldots,12\},+)$, the set of numbers on the clock face with "clock addition."
4. $(\mathcal{P}(A),\cap)$, the set of subsets of A with intersection.

A **binary operation** $\cdot$ on a set S may be regarded as a function $\cdot: S \times S \rightarrow S$ on the Cartesian product of S with itself into S, as was observed in Chapter 9. Thus, for example, $2 + 3 = 5$ may be expressed in functional notation as $+ (2,3) = 5$.

> ***11.1 Definition:*** Let A be a nonempty set. A **binary operation** ρ on A is a function $\rho: A \times A \rightarrow B$. A is said to be **closed** under ρ iff $B \subseteq A$.

The term "(binary) operation on a set" is often taken to automatically imply the property of closure. Here it is regarded as a separate property which may, but need not, be possessed by an arbitrary binary operation.

For example, the set $\mathbb{P}$ of positive integers is closed under ordinary addition, since the sum of two positive integers is itself a positive integer. Now, let D be the set of odd positive integers; then $D \subseteq \mathbb{P}$, and thus addition is still defined for D. (In fact $+$: $D \times D \rightarrow \mathbb{P}$ is a restriction of $+$: $\mathbb{P} \times \mathbb{P} \rightarrow \mathbb{P}$ to D.) Yet D is *not closed* under addition, since the sum of two odd integers is even and thus not in D.

Intuitively, the interpretation of "closed" is: when a set S is closed under an operation ρ, it is not possible to escape the set S using ρ; that is, starting with two elements of S, their ρ-combination must also be in S. When S is not closed under ρ, there is an "opening in S" through which ρ may lead. Thus, $3 + 5 = 8$ shows that addition leads "through a hole" in the set D of odd positive integers, since $8 \notin D$.

Closure usually comes into play when a *subsystem* of the original system is considered, as the original operation is usually defined on the original set so that the latter is closed under the former. The following are examples concerning closure, where the operations are often defined on a larger set than the one mentioned.

EXAMPLE 1:
$\mathbb{P}$ is closed under addition and multiplication, but not under subtraction ($2 - 3 \notin \mathbb{P}$) or division ($2/3 \notin \mathbb{P}$).

EXAMPLE 2:
The set E of even integers is closed under addition, subtraction, and multiplication. It is not closed under division ($6/2 = 3 \notin E$, and $6/0$ is not defined altogether).

EXAMPLE 3:
The set $\mathbb{R}$ of real numbers is closed under addition, subtraction, multiplication, and exponentiation. It is not closed under division (division by zero is excluded) or under root extraction ($\sqrt{-1} \notin \mathbb{R}$). On the other hand, $\mathbb{R} - \{0\}$ is closed under division and the set of positive real numbers is closed under root extraction.

11.2 THE (CAYLEY) OPERATION TABLE

Examples need not be familiar ones. In fact, one often encounters, or invents, "abstract" examples—an arbitrary set (usually finite and not too large) on which an operation is defined by exhaustive listing, such as a table called a **table of operation,** or **(Cayley) multiplication table** even though the operation is not the ordinary multiplication at all. (The reason for use of "multiplication" is perhaps the frequent use of the simple symbol dot ($\cdot$) as the operation symbol.)

In constructing such an example, one need only line up the elements of a set S in the **label column** on the left of the table and in the **label row** at the top of the table, and then arbitrarily fill the body of the table with entries. If the entries all come from S, the table defines an operation for which S is closed.

We do so with the set $S = \{a,b,c,d\}$ in Figure 11.1. From the table we may extract the image of any ordered pair of elements of S under dot. For example, $c \cdot b = d$ is taken from row c and column b, while $b \cdot c = c$ is taken from row b and column c. The convention is that, in looking up the definition of $x \cdot y$, the left element, x, labels a row while the second, y, labels a column.

$\cdot$	a	b	c	d
a	c	d	b	b
b	a	b	c	a
c	c	d	a	c
d	a	a	b	a

Operation table for $\cdot$ on S

FIGURE 11.1

Since no care was taken in assigning the "results" as entries in the table, it should not be expected to possess special properties. In fact, the closure of S under $\cdot$ was intended but a (nonempty) subset of S is not likely to be closed under (the restriction of) the same operation. For instance, $\{b,c\}$ is not closed under $\cdot$, since elements other than b and c appear in the intersection of rows b and c and columns b and c:

$$c \cdot b = d \notin \{b,c\}, \text{ and } c \cdot c = a \notin \{b,c\}.$$

11.3 ASSOCIATIVITY

If we select three elements at random, and perform the dot operation twice, we may get different results depending on which dot is performed first. For example, if the first dot in $b \cdot a \cdot b$ (Figure 11.1) is performed first, we get

$$(b \cdot a) \cdot b = a \cdot b = d,$$

while performing the second dot first yields

$$b \cdot (a \cdot b) = b \cdot d = a.$$

Of course, there may be triples of elements for which the *order of operations* does not matter (see $a \cdot b \cdot c$ and $b \cdot c \cdot d$, for example); however, as long as one example such as $(b \cdot a) \cdot b \neq b \cdot (a \cdot b)$ exists, $\cdot$ is not an associative operation.

> **11.2 Definition:**   Let S be a set and let $\cdot : S \times S \rightarrow S$ be a binary operation on S. Then $\cdot$ is **associative** iff, $\forall\ a,b,c \in S$,
>
> $$(a \cdot b) \cdot c = a \cdot (b \cdot c).$$

It is usually quite difficult to check whether the operation in an unfamiliar system is associative, unless one finds a counterexample—a case showing that it is not. Most familiar systems include associative operations:

EXAMPLE 1:
Addition and multiplication are associative on $\mathbb{R}$:

$$(a + b) + c = a + (b + c) \quad \text{and} \quad (ab)c = a(bc), \forall\ a,b,c \in \mathbb{R}.$$

On the other hand, the order of subtraction is crucial: $(10 - 5) - 2 = 5 - 2 = 3$, while $10 - (5 - 2) = 10 - 3 = 7$. Similarly with division, $(12/3)/2 = 4/2 = 2$, while $12/(3/2) = 12/1.5 = 8$. So neither subtraction nor division are associative operations.

EXAMPLE 2:
On the set $\mathcal{P}(A)$ of all subsets of A, union and intersection are associative: $(B \cup C) \cup D = B \cup (C \cup D)$ and $(B \cap C) \cap D = B \cap (C \cap D)$, $\forall\ B,C,D \in \mathcal{P}(A)$; but relative complementation is not: $(B - C) - D$ need not equal $B - (C - D)$ as is easy to verify.

It is always possible to test an operation on a finite set for associativity by exhaustion; i.e. checking systematically for each triple a,b,c (with duplication) whether $(a \cdot b) \cdot c = a \cdot (b \cdot c)$. However, this is often a horrendous task and is to be avoided whenever possible. Techniques for checking associativity, which improve upon exhaustion, do exist but even those are too involved for this presentation. We will avoid the test whenever we can, inviting the reader to accept our word for the associativity of an operation. We do, however, attempt to rely on devices which guarantee associativity in producing, and sometimes checking, associative operations. Two such devices are *concatenation* and *functional composition*. (Concatenation is reintroduced later in this section.)

Theorem 9.4, above, states without proof that *functional composition is associative*: Let $f : A \rightarrow B$, $g : B \rightarrow C$, and $h : C \rightarrow D$ be functions. Then,

$$(h \circ g) \circ f = h \circ (g \circ f).$$

The reasoning is quite simple from the definition of composition:

$$\forall x \in A, ((h \circ g) \circ f)(x) = (h \circ g)(f(x)) = h(g(f(x))),$$

while

$$(h \circ (g \circ f))(x) = h((g \circ f)(x)) = h(g(f(x)))$$

—the same result in both cases.

Armed with that knowledge, if we can express the repetition of an operation as functional composition (not always possible), we have associativity built in. (We do so for permutations in Section 12.8.)

11.4 SEMIGROUPS

Associativity is very convenient for obvious reasons. Consequently, a system with a set closed under an associative operation carries a special name.

> ***11.3 Definition:*** Let S be a nonempty set, and let $\cdot : S \rightarrow S$ be a binary operation (under which S is closed). Then the pair $(S, \cdot)$ is called a **semigroup** iff $\cdot$ is associative.

In other words, a semigroup is a set with a binary operation which obeys closure and associativity. Thus, all the aforementioned examples with closure and associativity are semigroups. Other examples of semigroups are:

1. The set B^A of all functions $f : A \rightarrow B$ on A to B, under functional composition.

2. The set of all permutations on a set (the operation is again composition).

3. All examples of groups (below) are also examples of semigroups.

11.5 THE FREE SEMIGROUP ON A SET

Consider the set $\{a,b\}$ of two elements, called "letters" here, as a "generating set." We are interested in the set of all "words" which can be generated from $\{a,b\}$ by **concatenation,** i.e. finite sequences of letters from $\{a,b\}$, which are obtained by placing a letter next to (on the right of) the previous letter.[1] Starting with the words with one letter and progressing to longer ones, the first elements (words) of our set, in lexicographic order, are:

$$a, b, aa, ab, ba, bb, aaa, aab, aba,$$
$$abb, baa, bab, bba, bbb, aaaa, \text{etc.}$$

This set is clearly infinite, since there is no longest member in it. (We can always concatenate another letter to make longer a word assumed to be longest.) We concatenate any two words into one in the obvious way:

$$(abb)(baab) = abbbaab.$$

Two elements of this set are equal iff they are *identical,* i.e. they look alike. The associativity of concatenation is obvious from its definition:

$$((a)(b))(c) = (ab)(c) = abc$$
while $(a)((b)(c)) = (a)(bc) = abc.$

Thus, *concatenation is associative.*

Since concatenation of two words over $\{a,b\}$ results in a

---

[1]See Definition 4.11 in Section 4.10. For more details on concatenation and strings, see Sections 4.10 and 4.16.

word over {*a,b*}, the set of words over {*a,b*} is closed. Consequently, we have a semigroup.

This is a special kind of semigroup, because there are no
relations defined on its set except for the identity; i.e. two
elements $x_1 x_2 \ldots x_k$ and $y_1 y_2 \ldots y_n$ are equal iff $k = n$ and $x_i =
y_i$ for each $i \in \{1, \ldots , k\}$. In other words, *the only equality
permitted is the identity*. Such a structure is said to be *free* of
relations and the semigroup too, is called free.

S^+

> **11.4 Definition:**   Let S be a nonempty set, and let S^+
> denote the set of all finite nonempty sequences of elements of
> S. Then S^+, with the operation of concatenation, is the **free
> semigroup** on S (**free semigroup generated by** S).[1]

The reader should be cautioned that a semigroup is often
written as just the set, leaving the operation understood but not
stated. Since there is only one operation defined on the set of a
semigroup, the context is usually sufficient to prevent confusion. In that manner, S^+ of the definition may serve both for
the free semigroup on S and as its set of elements.

11.6 SEMIGROUPS DEFINED BY RELATIONS

A table of an "abstract" finite semigroup can be
constructed using the construction of the preceding example,
but with the addition of some, arbitrary, relations on the
elements; this, of course, will result in a semigroup which is not
free. Nonetheless, it will be a semigroup, because concatena
tion is still associative (same reasoning applies) even with the
added relations.

For example, let us decide to "turn back" words before
they get too big by equating, say, $aaa = a$, $ba = ab$, and $bb = a$.
Let us adopt the *convention* that the *exponent on a letter
indicates the number of consecutive repetitions of the letter*, so
that $aaa = a^3$ and $bb = b^2$. The relations we use to modify
concatenation, and which we hope will yield just a finite
number of distinct elements, are then written as:

$$a^3 = a, \ ba = ab, \ b^2 = a.$$

To obtain successive elements of the set of words over {*a,b*},
modified by the relations, we start with the elements *a* and *b*,

---

[1]The reader may recognize the set S^+ as the positive closure of S from Section
4.16.

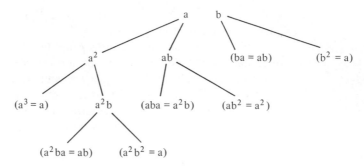

Generating words on {a,b} with relations

FIGURE 11.2

and successively concatenate each word (in its first encountered name) with both *a* and *b*, checking each time with the three relations whether the resulting word is another name to a previously generated element.

Thus, *a* and *b* are new elements. Using *a* as a starting point, we obtain $aa = a^2$ and ab, both of which are new. Using the second element as the starting point, we obtain $ba = ab$ and $bb = b^2 = a$, both of which are not new, and therefore may be parenthesized in Figure 11.2 and abandoned. We continue to "develop" in this fashion the original names of words, until (we hope) we reach a point at which no new words can be produced, because all concatenations with existing elements produce existing elements.

The computation displayed in Figure 11.2 may be easily used to construct the operation table of the semigroup we just created. The names which appear in the label column and label row are those which have not been parenthesized in Figure 11.2. The entries in the body of the table may be computed step by step from Figure 11.2.

·	a	b	a^2	ab	a^2b
a	a^2	ab	a	a^2b	ab
b	ab	a	a^2b	a^2	a
a^2	a	a^2b	a^2	ab	a^2b
ab	a^2b	a^2	ab	a	a^2
a^2b	ab	a	a^2b	a^2	a

Operation table of the semigroup

FIGURE 11.3

For example, the entry in row b and column a^2b is computed as follows:

$$
\begin{aligned}
(b)(a^2b) &= (ba)(ab) &\text{(associativity)}\\
&= (ab)(ab) &(ba = ab)\\
&= (a)(ba)(b) &\text{(associativity)}\\
&= (a)(ab)(b) &(ba = ab)\\
&= (a^2)(b^2) &\text{(associativity)}\\
&= (a^2)(a) &(b^2 = a)\\
&= a^3 &\text{(associativity)}\\
&= a &(a^3 = a)
\end{aligned}
$$

This computation may be cut short by use of previous entries in the table and Figure 11.3. For example, by the time we compute $(b)(a^2b)$, we already have $(b)(a^2) = a^2b$ and thus we could write:

$$
\begin{aligned}
(b)(a^2b) &= (ba^2)(b) &\text{(associativity)}\\
&= (a^2b)(b) &\text{(table entry)}\\
&= (a^2b^2) &\text{(associativity)}\\
&= a &\text{(Figure 11.2)}
\end{aligned}
$$

Two points of caution: the relations which modify concatenation may not always yield a finite set of words; even when they do, that set may be quite large. Also, care should be exercised not to introduce "hidden relations" which cause seemingly distinct elements to be identified.

11.7 SUBSEMIGROUPS

For purposes of reference, let us denote the set of elements of the semigroup in Figure 11.3 by T; i.e. $T = \{a,b,a^2,ab,a^2b\}$. Although it may not be easy to detect, there are three proper subsets of T on which the restriction of the concatenation operation is closed. They are $A = \{a^2\}$, $B = \{a,a^2\}$, and $C = \{a,a^2,ab,a^2b\}$. The closure in each case is easy to see when the rows and columns of elements not in the subset are deleted from the operation table, as was done in Figure 11.4. There, the only elements which appear in the body of the table are in the label rows and columns. (The dot appearing in each case should be taken as the restriction of the concatenation on T to the corresponding set.)

·	a²
a²	a²

(*i*) (A, ·)

·	a	a²
a	a²	a
a²	a	a²

(*ii*) (B, ·)

·	a	a²	ab	a²b
a	a²	a	a²b	ab
a²	a	a²	ab	a²b
ab	a²b	ab	a	a²
a²b	ab	a²b	a²	a

(*iii*) (C, ·)

Subsemigroups of (T, ·)

FIGURE 11.4

In each of the three cases, the restriction of the semigroup operation on the respective subset is still associative. The reason is that nothing has changed in the definition of the operation, except that now a smaller selection of elements has to satisfy the associative law; but these elements are among those which satisfied this law to begin with.

We then have three subsets of T, each closed with respect to an associative operation; this qualifies them as semigroups, according to Definition 11.3. Moreover, the semigroup operations are defined *identically* to the concatenation on T *wherever they are defined*, i.e., they are restrictions of the original semigroup operation. That makes them subsemigroups of $(T, ·)$.

11.5 Definition: Let $(T, ·)$ be a semigroup. Then $(S, \#)$ is a **subsemigroup** of $(T, ·)$ iff

(i) $S \subseteq T$, and
(ii) $\#$ is the restriction of $·$ to S; i.e., $\forall x, y \in S, \ x\#y = x·y$.

Two remarks may be helpful here. It is possible to define an operation on the same subset in a different way. The result may still be a semigroup, but it will not be a *subsemigroup* of the original.

The restriction of the semigroup operation to *any* subset of T is still going to be associative. If the subsystem is not a subsemigroup, that is because the subset is not *closed* under the (restricted) operation.

The reader should find it easy to think of examples of

subsemigroups of the semigroups mentioned above. For instance, the semigroup $(E,+)$ of the even positive integers with addition is a subsemigroup of $(\mathbb{P},+)$, the positive integers with addition.

11.8 IDENTITY

To illustrate another point, and at the same time to provide an additional example of the technique, we impose the relations

$$a^2 = a,\ ab = ba = b,\ ac = ca = c,\ c^2 = bc,\ b^2 = cb = a$$

on the set $\{a,b,c\}$, and obtain the semigroup in Figure 11.5(i). Part (ii) of the figure is a convenient translation of the table, obtained by replacing a,b,c, and bc by 1,2,3, and 4, respectively, and $ by # whenever they appear. (The two representations are **isomorphic** when one results from the other by one-to-one label trading.)

$	a	b	c	bc
a	a	b	c	bc
b	b	a	bc	c
c	c	a	bc	c
bc	bc	b	c	bc

#	1	2	3	4
1	1	2	3	4
2	2	1	4	3
3	3	1	4	3
4	4	2	3	4

(i)   (ii)

Isomorphic representations of a semigroup with identity

FIGURE 11.5

Using the representation in part (ii), the element 1 is seen to not change any element with which it is combined by the #-operation, whether on the left or on the right. For example, it can be seen from the table entries that $1\#3 = 3$ and $3\#1 = 3$. Such an element, *if it exists,* is called an **identity.** (Identity may be defined in a more general setting, although here we concentrate on semigroups.)

11.6 Definition: Let $(S,\cdot)$ be a semigroup. An element $i \in S$ is an **identity** (of the semigroup) iff,
$$\forall s \in S,\ s \cdot i = i \cdot s = s.$$

Not every semigroup has an identity element. $(T,\cdot)$ of Figure 11.3 does not have an identity, although each of its proper subsemigroups of Figure 11.4 does—a^2.

An identity should not be mistaken for an element which preserves multiples on one side and not on the other. We obtain the semigroup of Figure 11.6 on $\{a,b\}$ by imposing the relations $a^2 = a$, $b^2 = ab$, and $ba = b$.

*	a	b	ab
a	a	ab	ab
b	b	ab	ab
ab	ab	ab	ab

A semigroup with a right identity

FIGURE 11.6

This semigroup has the element a as a **right identity,** because $x*a = x$, for every x. However, a is not a **left identity** since $a*b = ab \neq b$. As a result, a is not an identity, as the latter must be both a left- and a right-identity.

When a semigroup has an identity element, no other element can serve as another identity.

11.7 Theorem: The identity of a semigroup (if it exists) is unique.

11.9 MONOIDS

A semigroup which does not have an identity may be supplied with one quite easily. For instance, we could add the "fictitious element" e to the semigroup of Figure 11.6, labeling a new row and a new column, making sure that it preserves each element with which it interacts. We do so in Figure 11.7, by extending the operation $*$ to the enlarged domain.

A semigroup with an identity is called a **monoid.**

11.8 Definition: Let $(S,\#)$ be a semigroup. Then $(S,\#)$ is a **monoid** iff S has an identity element under $\#$.

Just as we obtained the free semigroup on a set, we can use the same process of concatenation (with "equality" meaning

*	e	a	b	ab
e	e	a	b	ab
a	a	a	ab	ab
b	b	b	ab	ab
ab	ab	ab	ab	ab

A monoid for Figure 11.6

FIGURE 11.7

"identity") to produce the **free monoid** on the set, provided we could find an identity that would not introduce new equality relations. For that purpose we use the artificial construct of a *word with no letters*, denoted by λ; when it is concatenated on either side with *any* word, no letters are added and the word remains unchanged; thus, λ is the identity.

λ

S^*

> **11.9 Definition:** Let S be a set, and let S^* be the set of all sequences of finite length (including zero length) of letters from S. Then, S with the operation of concatenation is the **free monoid** on S. The **empty word,** denoted λ, is the sequence with no letters and of zero length, and concatenation with it is defined by, $\lambda x = x\lambda = x, \forall x \in S^*$, i.e., λ is the identity of the free monoid on S.

The concept of a **submonoid** should be fairly obvious at this point. Its set must be a subset of the monoid, it must contain the identity (the same one as in the original monoid), and the operation must be the restriction of the monoid operation to the (possibly) smaller domain.

A monoid is still a semigroup. So a submonoid of the monoid $(S,\#)$ is a subsemigroup of $(S,\#)$ which still has in it the identity of $(S,\#)$.

11.10 INVERSES

We now consider another property which may, but need not, be possessed by monoids. The monoid in Figure 11.8 has a as identity. Note that in the body of the table, a appears exactly once in each row and in each column. (in fact, this is true for every element.) For each appearance of a in the table, then, the labels of the row and column which intersect at the identity display the following property:

$$c \cdot b = a, \quad b \cdot c = a, \quad a \cdot a = a.$$

·	a	b	c
a	a	b	c
b	b	c	a
c	c	a	b

A group

FIGURE 11.8

In general, if the identity i appears in row x and column y, then $x \cdot y = i$. In fact, whenever that is the case, and it is also true that $y \cdot x = i$, the two elements x and y are each other's **inverses** and those, too, are unique.

> ***11.10 Definition:***   Let $(S, \cdot)$ be a monoid, with identity i, and let $x \in S$. Then an element $y \in S$ is the **inverse** of x iff $x \cdot y = y \cdot x = i$.

11.11 GROUPS

When a monoid has the property possessed by the monoid of Figure 11.8, to wit, *every element has an inverse in the monoid*, it is called a **group.** This requirement on each element that it possess an inverse turns out to be quite demanding, and as a result groups have many strong and interesting properties.

> ***11.11 Definition:***   A monoid $(S, \cdot)$ is a **group** iff, for every element x of S there exists an inverse of x in S. The inverse of x is denoted by x^{-1}.

The following are examples of monoids and groups.

EXAMPLE 1:
The set of nonnegative integers with addition is a monoid, with 0 as the identity, but it is not a group, since the inverse of 5 under addition is not in the set. To find the inverse, we must ask "what number when added with 5 gives the identity 0?" The answer of course is -5, which is not in the set. On the other hand, the set of all integers under addition is a group.

EXAMPLE 2:

The face of the clock with clock addition is a group. The identity is 12 and pairs whose sum is 12 are pairs of inverses: 2 and 10, 9 and 3, 6 and 6, etc.

EXAMPLE 3:

The nonzero rational numbers with multiplication form a group, with 1 as the identity and the reciprocal $1/x$ of each nonzero rational x as its inverse, since $x \cdot (1/x) = (1/x) \cdot x = 1$. On the other hand, the entire set of rationals with multiplication is a monoid but not a group, since 0 has no inverse: there exists no such number x so that $0 \cdot x = 1$.

11.12 SUMMARY OF AXIOMS FOR SEMIGROUPS, MONOIDS AND GROUPS

We summarize with the chart of Figure 11.9, in which the properties possessed by the set S and the binary operation $\#$ on S are matched with the structure they define.

The reader may have noticed that in the group of Figure 11.8, the order of elements does not matter; that is, $x \cdot y = y \cdot x$, for every x and y. This is *not* a property of all groups. A group which has this property is called **commutative** or **abelian.**

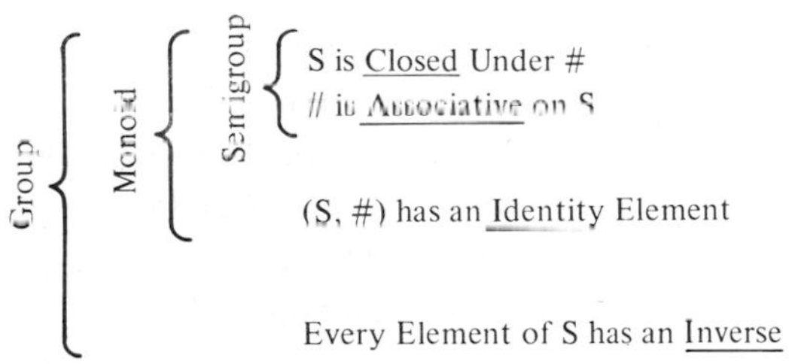

Semigroup, monoids, and group

FIGURE 11.9

12

Algebraic Concepts— Properties of Groups

12.1 DEFINITIONS AND SIMPLE RESULTS

Group theory is an extensively studied field. For the reader who needs to consult the fundamentals, we devote this chapter to a brief treatment of group properties that is hardly more than a list of definitions and elementary results.

We recall from Definition 11.11 that a **group** is a set G with a binary operation $\cdot$ defined on G so that

(i) G is closed under $\cdot$;

(ii) $\cdot$ is associative;

(iii) $(G, \cdot)$ has an identity element;

(iv) every $x \in G$ has an inverse in G under $\cdot$

The group $(G, \cdot)$ is often denoted just by G, without mention of the group operation.

$\cdot$	i	a	b	c	d	e
i	i	a	b	c	d	e
a	a	i	c	b	e	d
b	b	d	e	a	c	i
c	c	e	d	i	b	a
d	d	b	a	e	i	c
e	e	c	i	d	a	b

x	Inverse of x
i	i
a	a
b	e
c	c
d	d
e	b

A Group $(G, \cdot)$ | List of Inverses

FIGURE 12.1

We use as an example the group G whose operation table is shown in Figure 12.1. The reader is invited to find specific instances of the definitions and results in this example.

12.1 Theorem: Let $(G, \cdot)$ be a group. Then,

(i) the identity of G is unique;
(ii) every $x \in G$ has a *unique* inverse x^{-1} in G;
(iii) for every $x \in G$, $(x^{-1})^{-1} = x$;
(iv) for all $x,y \in G$, $(x \cdot y)^{-1} = y^{-1} \cdot x^{-1}$.

Part (iii) suggests that elements come in inverse pairs (e.g., b and e in the example). Part (iv) asserts that *the inverse of the product is the product of the inverses* (e.g., $c \cdot d = b$ and $(c \cdot d)^{-1} = b^{-1} = e$ while $d^{-1} \cdot c^{-1} = d \cdot c = e$).

12.2 Theorem: Let G be a group and let $a,b \in G$. Then the equations

$$a \cdot x = b \text{ and } y \cdot a = b$$

have unique solutions in G. The solutions are

$$x = a^{-1} \cdot b \text{ and } y = b \cdot a^{-1}.$$

12.3 Theorem: Any group G obeys the two cancellation laws:

$$a \cdot x = h \cdot x \Rightarrow a = b,$$

and

$$x \cdot a = x \cdot b \Rightarrow a = b.$$

12.2 ABELIAN (COMMUTATIVE) GROUPS

Recall that a group G is **abelian (commutative)** iff, $\forall a,b \in G$, $a \cdot b = b \cdot a$. We also adopt the convention that $a^0 =$ identity, $a^1 = a$, $a^2 = a \cdot a$, $a^3 = a \cdot a \cdot a$, etc.

We could easily reason out what, say, a^{-3} should denote, if the laws of exponents were still to apply, since then the inverse of a^{-3} should be a^3 so that $a^{-3} \cdot a^3 = a^{-3+3} = a^0 =$ identity. Consequently,

$$a^{-n} = (a^n)^{-1},$$

for all integers n and every element a in a group.

With this notation in hand, we have the following.

> **12.4 Theorem:**   Let G be an abelian group. Then, for all $a,b \in G$ and for all integers n,
>
> $$(a \cdot b)^n = a^n \cdot b^n.$$

Our example is not an abelian group and thus we can display an instance from it that does not obey the equality in the theorem. We have $b \cdot c = a$ so that $(b \cdot c)^2 = a^2 = i$, but $b^2 \cdot c^2 = e \cdot i = e$.

12.3 THE ORDER OF A FINITE GROUP

The number of elements of a finite group is called its order. The order of a group is a very important ingredient in the ensuing development.

$o(G)$

> **12.5 Definition:**   Let G be a finite group. Then the **order** of G is the number $\#(G)$ of elements of G and is denoted by $o(G)$.

In the example of Figure 12.1, $o(G) = 6$.

12.4 SUBGROUPS

The definition of a **subgroup** is similar to that of a subsemigroup and submonoid in that it is a subset which obeys all the group laws. In the following definition we call the operation and its restriction by the same name, since no confusion arises.

$<$

> **12.6 Definition:**   Let $(G, \cdot)$ be a finite group. Then, $(H, \cdot)$ is a **subgroup** of $(G, \cdot)$, written $H < G$, iff $H \subseteq G$ and $(H, \cdot)$ is itself a group. H is a **proper subgroup** of G iff H as a set is a proper subset of G.

Clearly, every subgroup must contain the identity element. Often the **identity subgroup,** whose only element is the identity,

·	i
i	i

·	i	a
i	i	a
a	a	i

·	i	c
i	i	c
c	c	i

·	i	d
i	i	d
d	d	i

·	i	b	e
i	i	b	e
b	b	e	i
e	e	i	b

A   B   C   D   E

The proper subgroups of $(G, \cdot)$

FIGURE 12.2

is not considered a *proper* subgroup.

The group of Figure 12.1 has the proper subgroups shown in Figure 12.2.

Note that $o(A) = 1, o(B) = o(C) = o(D) = 2$, and $o(E) = 3$. These facts are significant and we shall return to them.

The properties which must be inherited by a subgroup are closure, the presence of the identity, and the presence of an inverse for each element. These can be assured by the following conditions.

12.7 Theorem: A nonempty subset H of a group G is a subgroup of G iff

(i) if $a,b \in H$ then $a \cdot b \in H$, and
(ii) if $a \in H$ then $a^{-1} \in H$.

12.5 THE ORDER OF AN ELEMENT OF A GROUP

For *finite* groups, the conditions of Theorem 12.7 may be simplified further for the following reason. Let G be a finite group and let $a \in G$. Then each member of the infinite list $a, a^2, a^3, \ldots$ of powers of an element a of G is a name for some element of G (by closure). But there are only finitely many elements in G. Consequently, for some positive integers n and k, such that $n > k$, it must be that $a^n = a^k$. Therefore, multiplying both sides on the right by a^{-k}, we get

$$a^n = a^k$$
$$\Rightarrow a^n a^{-k} = a^k a^{-k}$$
$$\Rightarrow a^{n-k} = a^{k-k} = a^0 = i,$$

where i is the identity of G. We thus conclude that for each element a of a finite group there is a positive integer n so that a^n is the identity. The smallest such positive integer is called the **order** of the element a and is denoted by $o(a)$. (Note that we use the same notation for the order of an element and the order of a group, but no confusion should arise. Also note that, if the group is infinite—has an infinite set G—then the infinite succession $a, a^2, a^3, \ldots$ need not have any repetition and all elements may be distinct. Then $o(a)$ is infinite.)

$o(a)$

> **12.8 Definition:**   Let G be a group and let a be an element so that, for some positive integer n, a^n equals the identity. Then the least such n is called the **order** of a and is denoted by $o(a)$.

In a finite group, each element a generates a **cyclic** subgroup whose elements are $a, a^2, \ldots, a^{o(a)}$, and therefore its order equals $o(a)$.

The orders of elements in our example are: $o(i) = 1$, $o(a) = o(c) = o(d) = 2$, $o(b) = o(e) = 3$. The correspondence between these orders and the orders of the subgroups in which they appear is not a coincidence.

It is a simple matter to realize that, *in a finite group, the inverse of each element* a *is some power* a^k *of itself:* we have $a^{o(a)} = i$ and thus $a \cdot a^{o(a)-1} = i$, so that $a^{o(a)-1} = a^{-1}$ unless $a = i$, in which case $a^{-1} = a$. It is now possible to hone down Theorem 12.7 for the case of finite groups, since its part (ii) is automatic from part (i) in a finite group.

> **12.9 Corollary:**   A nonempty subset H of a finite group G is a subgroup of G iff $a, b \in H \Rightarrow a \cdot b \in H$.

12.6 COSETS

There are several simple consequences of the definition of the order $o(a)$ of a group element a. However, they may be easier to perceive after we introduce the notion of a **coset** of a subgroup. We use the subgroup E (Figure 12.2) of the group G (Figure 12.1) and copy both for the reader's convenience in Figure 12.3.

·	i	a	b	c	d	e
i	i	a	b	c	d	e
a	a	i	c	b	e	d
b	b	d	e	a	c	i
c	c	e	d	i	b	a
d	d	b	a	e	i	c
e	e	c	i	d	a	b

·	i	b	e
i	i	b	e
b	b	e	i
e	e	i	b

Group G   Subgroup E

FIGURE 12.3

Now, consider the products of d on the left with the elements of E. We get

$$d \cdot i = d, \quad d \cdot b = a, \quad d \cdot e = c.$$

We write $dE = \{dx: x \in E\} = \{d,a,c\}$, and call dE a **left coset** of E in G. We note that the three elements of dE must be distinct from each other, by Theorem 12.2. However, we would obtain the same set $\{a,c,d\}$, although in a different order (which does not matter), when we use a or c instead of d to create the left coset, i.e.

$$aE = \{a \cdot i, a \cdot b, a \cdot e\} = \{a,c,d\}, \text{ and}$$

$$cE = \{c \cdot i, c \cdot b, c \cdot e\} = \{c,d,a\}.$$

That is, $aE = cE = dE = \{a,c,d\}$. It should be clear from the process that *the size of the coset equals the order of the subgroup,* i.e. $\#(aE) = o(E)$

Had we multiplied E on the *right* by d, we would have got the **right coset**

$$Ed = \{x \cdot d: x \in E\} = \{d,c,a\}.$$

(It is a coincidence that we obtained the same elements for dE and Ed, and we return to it later in the section. When we try the same with the left coset aD and right coset Da, we obtain $aD = \{a \cdot i, a \cdot d\} = \{a,e\}$ and $Da = \{i \cdot a, d \cdot a\} = \{a,b\}$.)

Whatever element is used to multiply the subgroup, the result is a set whose size is $o(E)$. Moreover, *two left* (alt., right) *cosets are identical* ($aE = cE = dE$) *or disjoint,* as we see when we use the remaining three elements: $iE = bE = eE = E$, since i,b,e are already elements of E, and E is closed under $\cdot$. It might be instructive to observe the left cosets of D:

$$iD = dD = D = \{i,d\},$$
$$aD = eD = \{a,e\},$$
$$bD = cD = \{b,c\}.$$

Thus, we *partition* G by creating left cosets (alt., right cosets with a possibly different partition) into sets each of the size of the subgroup. Note that E is always a left (right) coset of itself; i.e., $aE = E, \forall a \in E$.

We summarize the previous discussion.

12.10 Definition: Let H be a subgroup of a group G, and let $a \in G$. Then, $aH = \{ah: h \in H\}$ is a **left coset** of H in G and $Ha = \{ha: h \in H\}$ is a **right coset** of H in G.

12.11 Theorem: Let G be a group and let H be a subgroup of G. Any two left (alt., right) cosets of H are identical or disjoint (have no element in common).

12.12 Theorem: Let G be a group and let H be a subgroup of G. Then there is a 1–1 correspondence between any two left (alt., right) cosets of H in G. If G is finite, all left (alt., right) cosets of H in G have the same size as $o(H)$.

A finite group G is then the disjoint union of its left cosets by a subgroup H, each coset of the same size as H. Then $o(G)$ must be a multiple of $o(H)$.

12.13 Lagrange's Theorem: Let G be a finite group and let $H < G$. Then $o(H)$ is a divisor of $o(G)$; i.e., $o(H) \mid o(G)$.

The reader may recall that, in the example of Figure 12.1, $o(G) = 6$, and that the orders of all proper subgroups of G are 1, 2, and 3—all divisors of 6.

12.14 Definition: Let G be a group, and let $H < G$. The **index** of H in G is the number of distinct left (alt., right) cosets of H in G. In case G is finite, the index of H in G is $o(G)/o(H)$.

⟨a⟩

Since any element a of a finite group G generates a finite (cyclic) subgroup ⟨a⟩ with elements $a, a^2, \ldots, a^{o(a)} = i$, the order $o(a)$ of a equals the order $o(\langle a \rangle)$ of the subgroup. Thus we have

12.15 Theorem: Let G be a finite group and let $a \in G$. Then $o(a)$ divides $o(G)$; i.e.,

$$o(a) \mid o(G).$$

Because of Theorem 12.15, $o(G)$ is a multiple of $o(a)$, for each $a \in G$; therefore the product of $o(G)$ copies of *any* element of G is the identity:

12.16 Theorem: Let G be a finite group and let $a \in G$ Then

$$a^{o(G)} = i.$$

12.7 NORMAL SUBGROUPS

In marshalling a few fundamental ideas from group theory, we briefly touch on **normal subgroups,** but only superficially.

When we found the left and right cosets of the subgroup E of G (Figure 12.3), we saw that $xE = Ex$ for each $x \in G$; yet this was not true for the subgroup D, since $aD = \{a,e\} \neq \{a,b\} = Da$. A subgroup such as E has special important properties, of which we mentioned but several.

The criterion $xE = Ex$, $\forall x \in G$, is equivalent to the condition $xEx^{-1} = E$, since multiplying both sides of $xE = Ex$ by x^{-1} on the right yields $xEx^{-1} = Exx^{-1} = E$. Another equivalent condition is the customary one used in the definition of a normal subgroup.

◁

12.17 Definition: A subgroup N of a group G is a **normal subgroup** of G, written $N \triangleleft G$, iff, for every $g \in G$ and every $n \in N$,

$$gng^{-1} \in N.$$

12.18 Corollary: Let G be a group and let $N < G$. Each of the following statements is equivalent to the others.

(a) N is a normal subgroup of G; i.e., $N \triangleleft G$.

12.18 Corollary—Continued

    (b) For every $g \in G$, $gNg^{-1} = g^{-1}Ng$.

    (c) Every left coset of N in G is a right coset of N in G.

    (d) The product $(xN)(yN) = \{xnym\colon n,m \in N\}$ of any two left cosets of N in G is again a left coset of N in G. (The same is true for right cosets of N in G.)

The idea in part (d) of Corollary 12.18 is an important one. It indicates that the set of cosets of a normal subgroup N is *closed under multiplication of cosets.*

G/N

12.19 Definition: Let G be a group and let $N \lhd G$. Then the **quotient** of G by N is the set of cosets of N in G, denoted by G/N. The **product** of two members A and B of G/N is $A \cdot B = \{a \cdot b\colon a \in A \text{ and } b \in B\}$.

The reader may find the details of the following theorem not too difficult to verify at this point.

12.20 Theorem: Let G be a group and let $N \lhd G$. Then,

    (i) G/N is closed under set product ($A,B \in G/N \Rightarrow AB \in G/N$).

    (ii) Set product is associative on G/N ($A,B,C \in G/N \Rightarrow (AB)C = A(BC)$).

    (iii) N is the identity for G/N under set product.

    (iv) Each member of G/N has an inverse in G/N. (Na^{-1} is the inverse of Na in G/N, for each $a \in G$.)

Thus, $(G/N,\cdot)$ is a group.

12.21 Definition: Let G be a group and let $N \lhd G$. Then the group $(G/N,\cdot)$ of cosets with set product is called the **quotient group** or **factor group** of G by N.

The following results derive from the above with little difficulty. (Assume G is finite when its size is considered.)

1. If $N \lhd G$, then $o(G/N) = o(G)/o(N)$.

2. A subgroup of index 2 is always normal.

3. If $N \lhd G$ and $H < G$ then $NH < G$.

4. If $N \lhd G$ and $M \lhd G$ then $(N \cap M) \lhd G$ and $NM \lhd G$.

5. Every subgroup of an abelian group is normal.

6. If H is the only subgroup of order $o(H)$ in a finite group G, then $H \lhd G$.

7. $NaNb = Nab$, $\forall a,b \in G$. (Reason: $NaNb = NNab = Nab$.)

12.8 PERMUTATION GROUPS

In the brief discussion of permutations in Section 10.4, they were introduced as monic and epic functions on a set to itself. Here we take a brief look at **groups of permutations** and, although much of what is said is true for permutations on infinite sets, we shall restrict our attention here to permutations on finite sets.

Since the particular labels of the elements are not important (the *rearrangement* is of importance), we are concerned then with the size of the set—the number of its elements. We S_n denote the **set of all permutations on n letters** by S_n. Each $x \in S_n$ is a monic and epic function $x:N \to N$, where N is a set of n elements, and therefore functional composition is a binary operation on S_n. (We shall denote the composition of permutations by a dot, which we shall omit when convenient.) For example, if $x = (ac)(b)$ and $y = (a)(bc)$ are elements of S_3, then $yx = y \cdot x$ is defined by,

$$(yx)(a) = y(x(a)) = y(c) = b,$$

$$(yx)(b) = y(x(b)) = y(b) = c,$$

$$(yx)(c) = y(x(c)) = y(a) = a.$$

In the standard notation, $yx = (abc)$, since yx maps a to b, b to c, and c to a.

The composition of two monic and epic functions is itself monic and epic, so that the above illustration is typical and the *composition of permutations is itself a permutation.* Hence S_n is *closed* under the operation.

As was remarked in Section 11.3, functional composition is always associative. Thus, $\cdot$ is *associative* on S_n.

The **identity permutation,** the one which maps each letter to itself, also acts as the identity on $(S_n, \cdot)$, since in combination with any other permutation, it leaves that permutation unchanged. Thus, $(S_n, \cdot)$ possesses an *identity* element.

Lastly, as was seen in the earlier discussion in Section 10.4, each permutation has an inverse in S_n. In fact, to obtain the inverse of a permutation in the notation employing a base row and an image row below it, e.g. $\begin{pmatrix} a\ b\ c \\ b\ c\ a \end{pmatrix}$, the inverse is obtained by reading the permutation in reverse, i.e., inverting the two rows, e.g. $\begin{pmatrix} b\ c\ a \\ a\ b\ c \end{pmatrix}$, and reordering the columns, i.e. $\begin{pmatrix} a\ b\ c \\ c\ a\ b \end{pmatrix}$. Thus, every member of S_n has an *inverse* in $(S_n, \cdot)$.

All requirements for a group are satisfied and hence $(S_n, \cdot)$ is a group, called the **symmetric group of order** n, or the **group of permutations** of n objects (letters).

Permutation groups and their subgroups are important in themselves, but they carry added importance because of the fact that *any group is essentially a subgroup of some permutation group.* We recall that two structures are **isomorphic** (more on that in Chapter 13) whenever one is obtained from the other by just relabeling.

> ### 12.22 *Cayley's Theorem:*   Every finite group is isomorphic to a subgroup of S_n for some n.
> (Where S_A is the symmetric group on an arbitrary set A, every group is isomorphic to a subgroup of S_A, for some A.)

13

Algebraic Concepts— Homomorphisms, Isomorphisms and Automorphisms

13.1 HOMOMORPHISMS

A **homomorphism** is a structure-preserving mapping[1]. It is more than a function $f: A \to B$ on one set to another; for f to be a homomorphism, there has got to be a *structure* (such as one or more binary operations) associated with the domain A, a like structure associated with the co-domain B, and the manner in which f assigns values in B to members of A must mirror the similarity of the two structures. In what follows, we make this precise.

Homomorphisms are a frequent and important tool in many areas, including automata and sequential machines, formal languages and complexity, optimization of programs, denotational semantics, data structures (data types as algebras), automatic compilation, and others. Homomorphisms are helpful in situations of loss of information, whether intentional or accidental, such as data transmission and compacting, image processing, even in computer design. Homomorphisms

---

[1]A preliminary treatment of functions which preserve properties is included in Section 10.5.

are a major tool in the study of groups and semigroups. Semigroup homomorphisms are the simplest of the lot and are helpful in understanding homomorphisms of more complex systems; therefore, we start with them.

13.2 HOMOMORPHISMS ON SEMIGROUPS

We then have two semigroups, $(S, \cdot)$ and $(T, \#)$, and a function f which maps the members of S to members of T. Let us trace two members of S, say a and b, in the illustration in Figure 13.1

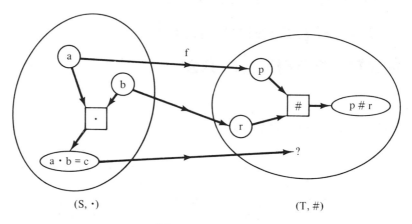

What is the f-image of a·b?

FIGURE 13.1

There are two things which happen to a and b: on the one hand, each is mapped by f to a member of T; and on the other, the binary operation $\cdot$ on S maps the pair (a,b) to some member of S. To be specific, we let

$$f(a) = p \text{ and } f(b) = r, \text{ in } T; \text{ and } a \cdot b = c, \text{ in } S.$$

Now, $f: S \to T$ maps c to some element of S. There is no reason why that image should be precisely $p \# r$. That is, why should $f(a) \# f(b)$ and $f(a \cdot b)$ be the same? And if they *happen* to equal, will this be true for *any* choice of a and b?

Unless the function f is defined very carefully (sometimes it cannot be done altogether), we have no reason to expect this perfect fit. What this would be asking is that, no matter which route we take, we always get the same result: given any a and b in S, *it does not matter* whether we *first* use on them the binary operation in S to get $a \cdot b$ and *then* map the result by f into T to get $f(a \cdot b)$, or we *first* map a and b separately into T to get $f(a)$ and $f(b)$, and *then* combine them in T by its binary operation #. In short, we would be demanding that,

$$\forall a,b \in S, f(a \cdot b) = f(a) \# f(b).$$

To see this more clearly, consider the two semigroups $(S, \cdot)$ and $(T, \#)$ of Figure 13.2.

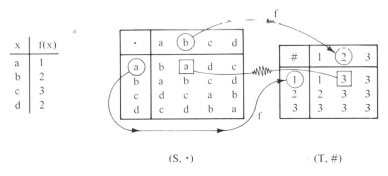

(S, ·)   (T, #)

Not a homomorphism

FIGURE 13.2

The function f, whose definition is shown on the left of the figure, maps a to 1 and b to 2, but not $a \cdot b$ to 1#2, as the former is $f(a \cdot b) = f(a) = 1$ while the latter is $f(a) \# f(b) = 1 \# 2 = 3$.

Actually, in this case it is possible to define a function g: $S \to T$ which preserves the operation · on S. We simply let g map every element of S to 3. Then, for any two elements x and y of S (the reader should select a particular pair here),

$$g(x \cdot y) = 3, \text{ since } x \cdot y \in S,$$

and

$$g(x) \# g(y) = 3 \# 3 = 3.$$

Thus, g is a homomorphism.

We give less trivial examples of homomorphisms below. Again we denote by $\mathbb{P}$ the set of positive integers.

EXAMPLE 1:

From $(\mathbb{P},+)$, i.e. ordinary addition on the positive integers, into $(\{0,1\},\oplus)$, where $\oplus$ is the **symmetric difference**[1] defined by

$$0 \oplus 0 = 1 \oplus 1 = 0, 0 \oplus 1 = 1 \oplus 0 = 1.$$

The function f mapping even integers to 0 and odd integers to 1 is a homomorphism.

E.g., $f(22 + 3) = f(25) = 1 = 0 \oplus 1 = f(22) \oplus f(3)$.

EXAMPLE 2:

From $(\mathbb{P},\cdot)$ into $(\{0,1\},\cdot)$, i.e. ordinary multiplication on $\mathbb{P}$ and ordinary multiplication on $\{0,1\}$. The same function $f\colon \mathbb{P} \to \{0,1\}$ of example 1 is a homomorphism here.

EXAMPLE 3:

From $(\mathbb{P},+)$ to $(\mathbb{P},+)$ with the function $g\colon \mathbb{P} \to \mathbb{P}$ defined by $g(x) = 2x$. (Note that g is not epic, as the odd members of $\mathbb{P}$ are not in the range of g.) We show that g is a homomorphism by selecting any x and y from $\mathbb{P}$ and computing

$$f(x + y) = 2(x + y) = 2x + 2y = f(x) + f(y).$$

We now define a homomorphism for semigroups, but that same definition holds for monoids and groups, since they are semigroups.

13.1 Definition: Let $(S,\cdot)$ and $(T,\#)$ be semigroups and let $f\colon S \to T$. Then, f is a **homomorphism** iff,

$$\forall a,b \in S, f(a \cdot b) = f(a) \# f(b).$$

The image set $f(S)$ is called the **homomorphic image** of $(S,\cdot)$. If $(T,\#) = (S,\cdot)$, a homomorphism $f\colon S \to S$ is called an **endomorphism** and $f(S)$ is the **endomorphic image** of $(S,\cdot)$.

Since the homomorphic image of a semigroup $(S,\cdot)$ in a semigroup $(T,\#)$ is a subset of T, it inherits associativity from $(T,\#)$. Then if we verify closure for $f(S)$, the homomorphic image of a semigroup must be a semigroup. But any two members of $f(S)$ are f-images of some members of S and so, for say $a,b \in S$,

$$f(a) \# f(b) = f(a \cdot b),$$

---

[1]The symmetric difference is the subject of Section 4.5.

since f is a homomorphism. But $a \cdot b \in S$, hence $f(a) \# f(b) \in f(S)$ and we have closure.

> **13.2 Theorem:**   Let f be a homomorphism on a semigroup $(S, \cdot)$ to a semigroup $(T, \#)$. Then, $(f(S), \#)$ is a semigroup.

13.3 HOMOMORPHISMS ON MONOIDS

Let us see what happens to the identity i of a monoid $(S, \cdot)$ that is the domain of a semigroup homomorphism, $f: (S, \cdot) \rightarrow (T, \#)$. Let x be any element of the domain; then $f(x) = f(x \cdot i) = f(x) \# f(i)$ and $f(x) = f(i \cdot x) = f(i) \# f(x)$. It turns out, then, that the element $f(i) \in T$, to which f maps the identity of S, behaves like, and therefore is, the identity element of $(T, \#)$. (We discovered that $(T, \#)$ is a monoid!) We then have:

> **13.3 Theorem:**
>
> *(i)* The homomorphic image of a monoid is a monoid.
> *(ii)* A homomorphism on a monoid maps the identity of the domain to the identity of the range.

13.4 HOMOMORPHISMS ON GROUPS

Continuing in the same vein, what does a group-homomor phism $f: G \rightarrow H$ do to the *inverse* of an element? Consider an arbitrary element x of the domain group G. (We shall omit both operation symbols, we hope without causing confusion at this point.) Then, since f maps the identity i_G of G to the identity i_H of H, we have

$$f(x) f(x^{-1}) = f(xx^{-1}) = f(i_G) = i_H.$$

Since the inverse (when it exists in a monoid) is unique, $f(x^{-1})$ must be the inverse of $f(x)$; i.e. $f(x^{-1}) = f(x)^{-1}$. We thus have

> **13.4 Theorem:**
>
> *(i)* The homomorphic image of a group is a group.
> *(ii)* A group-homomorphism maps inverse pairs to inverse pairs.

K_f

The subset K_f of elements of G which a homomorphism $f: G \rightarrow H$ maps to the identity of H is called the **kernel** of f; this subset has interesting and important properties. Note that there may be two homomorphisms f and g on G to H, and the two kernels K_f and K_g will not be identical sets if $f \neq g$.

> **13.5 Definition:**   Let f be a group homomorphism on G into H, and let i_H be the identity of H. The **kernel** of f is defined by
>
> $$K_f = \{x \in G : f(x) = i_H\}.$$

[The discussion which follows, and which ends with Theorem 13.7, is too terse (and insufficiently illustrated) to serve successfully as a first exposure; it appears here mainly to provide the reader who already has a background in group theory with a brief summary of some basic results and reasoning.]

Recall from Section 12.7 that, when N is a *normal* subgroup of G, G/N is a group—the *quotient group* whose elements are the cosets of N in G. Now consider the *natural function f* which maps each element x to the coset in which it is a member: f can be defined as $f : G \rightarrow G/N$ such that $f(x) = Nx$, because the presence of the identity in the subgroup N assures that the product $ix = x$ is one of the members of Nx.

Now, the product of any two cosets Nx and Ny of N is

$$(Nx) \cdot (Ny) = N(xN)y = N(Nx)y = (NN)xy = Nxy,$$

i.e. itself a coset of N. (Note that $NN = N$ by Theorem 12.20(iii).) Thus,

$$f(x)f(y) = (Nx)(Ny) = Nxy = f(xy),$$

and f is a homomorphism.

Since every coset is the image of some member of G, f is epic; therefore we have the following.

> **13.6 Theorem:**   Let G be a group and let $N \lhd G$ (i.e., N is a normal subgroup of G). Define $f : G \rightarrow G/N$ by $f(x) = Nx$, $\forall x \in G$. Then f is a homomorphism of G onto G/N.

Now, what is the kernel K_f of this natural homomorphism? The identity of G/N is N, so we seek all members of G which f

maps to N (the identity coset of G/N). But, by the definition of f, these are precisely the members of N, so that $K_f = N$.

As it turns out, the kernel of a group homomorphism is always not just a subgroup of the domain, but a normal one.

>   ***13.7 Theorem:*** Let $f: G \to H$ be a group homomorphism with kernel K. Then,
>
> $$K \lhd G.$$

13.5 ISOMORPHISMS

It must be clear by now that a homomorphism need not be monic, nor does it need to be epic. We call a monic and epic homomorphism an **isomorphism,** although some authors require it only to be monic. (In that case what we call an isomorphism would have to be called an epic isomorphism.)

>   ***13.8 Definition:*** A semigroup (alt., monoid, group) homomorphism $f: S \to T$ is an **isomorphism** iff it is monic and epic. Two semigroups are said to be **isomorphic,** written $S \cong T$, iff there is an isomorphism on one onto the other.

$\cong$

It should be clear that if S is isomorphic to T then T is isomorphic to S; i.e., $f: S \to T$ is an isomorphism iff $f^{-1}: T \to S$ is an isomorphism.

It should also be clear that *isomorphism is an equivalence relation,* since it is reflexive ($G \cong G$), symmetric ($G \cong H \Rightarrow H \cong G$), and transitive ($G \cong H$ and $H \cong K \Rightarrow G \cong K$).

The concepts just mentioned are the subjects of entire areas in algebra, but we are approaching the limit of their usefulness in this book. So, we present two additional, fairly central results before moving on.

>   ***13.9 Theorem:*** A group homomorphism $f: G \to H$ with kernel K_f is an isomorphism iff $K_f = \{i_G\}$; i.e. if the only element in the kernel is the identity of G.

>   ***13.10 Theorem:*** Let $f: G \to H$ be a group homomorphism with kernel K. Then $G/K \cong H$.

Theorem 13.10 is a very important one in group theory, and has many far-reaching applications.

13.6 AUTOMORPHISMS

When the domain and range of a group isomorphism are the same, i.e. $f\colon G \to G$, it is called an **automorphism.** (The same is true for semigroups, monoids, automata, and other algebraic structures. However, here we focus on group automorphisms.)

> *13.11 Definition:* An **automorphism** of a group G is an isomorphism of G onto G.

$\mathcal{A}(G)$

All automorphisms of a group G have the same domain and range, and therefore functional composition forms a binary operation on their set. The **set of all automorphisms** on G is often denoted by $\mathcal{A}(G)$. This symbol also denotes the pair consisting of the set $\mathcal{A}(G)$ and functional composition as the binary operation on the set, which we shall take as assumed henceforth.

(The reader to whom this material is new is urged to go through the steps of a proof to the following theorem. Such a proof is quite straightforward.)

> *13.12 Theorem:* Let G be a group and let $\mathcal{A}(G)$ be the set of automorphisms of G. Then $\mathcal{A}(G)$ is a group.

Again, a great deal more than can be usefully presented here may be of interest. We offer just one additional result on the subject of automorphisms of a group. (The reader who has further interest is referred to the reading list.)

> *13.13 Theorem:* Let G be a group, and let $f\colon G \to G$ be an automorphism of G. Then f maps every element to an element of the same order, i.e.
>
> $$o(f(x)) = o(x), \forall\, x \in G.$$

13.7 CONGRUENCE RELATIONS; PARTITIONS WITH THE SUBSTITUTION PROPERTY

This chapter will not be complete without combining homomorphisms with equivalence relations to obtain **congruence relations.**

We recall[1] that an equivalence relation is induced on a set S just by partitioning S into blocks, which are disjoint subsets whose union is the set. The equivalence relation is expressed on two elements x and y by "x and y are in the same block."

Now, suppose that there is a binary operation defined on S. If we subdivide S indiscriminately into blocks, we should not expect the induced equivalence relation to carry through the binary operation.

We illustrate with the set $\{0,1,2,3\}$ and addition modulo 4, for which the table of operation is shown in Figure 13.3 (i). Now, part (ii) shows the result of the partition $\{(0,1),\{2,3\}\}$—block boundaries are crossed when sums by all combinations are found. In contrast, the partition $\{\{0,2\},\{1,3\}\}$ in part (iii) of the figure *preserves the operation*. It has the **substitution property:** Let "$\equiv$" denote "is in the same block as."

+	0	1	2	3
0	0	1	2	3
1	1	2	3	0
2	2	3	0	1
3	3	0	1	2

	$\{0,1\}$	$\{2,3\}$
$\{0,1\}$	$\{0,1,2\}$	$\{0,2,3\}$
$\{2,3\}$	$\{0,2,3\}$	$\{0,1,2\}$

·	$\{0,2\}$	$\{1,3\}$
$\{0,2\}$	$\{0,2\}$	$\{1,3\}$
$\{1,3\}$	$\{1,3\}$	$\{0,2\}$

(*i*) Addition (Mod 4)   (*ii*) Equivalence   (*iii*) Congruence

FIGURE 13.3

(E.g. in the partition of part (iii), $0 \equiv 2$ and $1 \equiv 3$.) Then, $x \equiv y$ and $u \equiv v$ together imply that $x + u \equiv y + v$, i.e. one member of the block may be *substituted* for another and the result will still be in the same resulting block.

Such a partition is said to possess the substitution property and is called a **substitution property** (abbreviated, **s.p.**) **partition.** The induced equivalence relation is called a **congruence relation.** It is an equivalence relation which preserves the operation. In other words, the binary operation on the set S induces a binary operation on the *partition,* one which combines two blocks to yield a block.

Moreover, the natural function f_π on S to the partition π (which maps each element of S to the block housing it) is a *homomorphism* in the case of a congruence relation and an s.p. partition. That is to say, it preserves the binary operation.

---

[1]See Sections 7.16–7.19 for a discussion of the topics: equivalence relations, equivalence classes, partitions and the substitution property.

In the example of Figure 13.3 (iii), the natural function f is defined by

$$f(0) = f(2) = \{0,2\}, f(1) = f(3) = \{1,3\}.$$

Now,

$$f(2 + 3) = f(1) = \{1,3\}, \text{ and}$$
$$f(2) \cdot f(3) = \{0,2\} \cdot \{1,3\} = \{1,3\},$$

so

$$f(2 + 3) = f(2) \cdot f(3).$$

It is easy to verify for the rest of the cases that the preceding computation is typical and that f is a homomorphism.

14

Binary Arithmetic and
Other Bases

14.1 POSITIONAL NOTATION

The key idea in manipulation of numbers in binary notation (i.e., base 2 instead of base 10) is still that of **positional
notation.** In our decimal system a digit of a number, say the
leftmost 3 of 4303, has a value determined in part by the
position (from the right) which it occupies in the number.
Thus, the rightmost 3 in 4303 represents three units, while the
leftmost 3 (looking identical) represents three hundred units.

The reason, of course, is that the first position from the
right in a number is reserved for the number of units, the
second for the number of tens, the third for the number of
hundreds, and so on. The succession, "units, tens, hundreds,
..." is actually the succession

$$10^0, 10^1, 10^2, \ldots$$

and is used just because our system has 10 as its base. If the
base were 2, instead, this succession would be

$$2^0, 2^1, 2^2, 2^3, \ldots$$

only it would apply in a number from right to left, as is shown
in Figure 14.1.

32's	16's	8's	4's	2's	1's	
... 2^5	2^4	2^3	2^2	2^1	2^0	◄——— Value of Position
1	1	0	1	1	0	◄——— How Many of Each

32 + 16 + 4 + 2 = 54 in binary notation

FIGURE 14.1

14.2 CONVERTING BINARY TO DECIMAL

Just as we used the digits $0, 1, \ldots, 9$ in base 10 (i.e., from 0 to one less than the base), we would use the digits $0, 1$ in base 2. The binary number 110110 in Figure 14.1 may be translated as

$$1 \cdot 32 + 1 \cdot 16 + 0 \cdot 8 + 1 \cdot 4 + 1 \cdot 2 + 0 \cdot 1$$
$$= 32 + 16 + 0 + 4 + 2 + 0 = 54.$$

(It is easier to start from the right, since the power of 2 need not be guessed then.)

To convert a binary integer to its decimal equivalent, add the powers of 2 corresponding to the positions of the 1's in the binary integer. (In higher bases than 2, the task is not quite as simple, as is shown in Section 14.9.)

14.1 Algorithm (Binary to Decimal): Given a positive **binary integer** (a positive integer represented in binary notation—base 2), to obtain its decimal equivalent (the same integer represented in decimal notation—base 10). Let the binary integer be represented as $a_n a_{n-1} \ldots a_1 a_0$, where n is a nonnegative integer.

Step 1. Initialize k to 0 and the decimal integer N to 0. (Do Step 2.)

Step 2. If $a_k = 1$, add 2^k to N. (Do Step 3.)

Step 3. If $k < n$, increment k by 1 and do Step 2. Otherwise, stop. ∎

When the binary number has both a whole part (an integer) and a fraction (the part to the right of the point), the whole part is computed separately, as illustrated above. To compute the fractional part, we must first establish the values of the positions to the right of the point. As is the case in decimal arithmetic, we simply continue the descending order of the exponents of 2, using negative integers: $2^{-1} = 1/2$, $2^{-2} = 1/4$, $2^{-3} = 1/8$, etc.

Let our binary number be 110110.101101. Then the fractional part may be computed as $1 \cdot 1/2 + 0 \cdot 1/4 + 1 \cdot 1/8 + 1 \cdot 1/16 + 0 \cdot 1/32 + 1 \cdot 1/64 = 45/64 = .703125$, as may be seen from Figure 14.2.

	$\frac{1}{2}$'s	$\frac{1}{4}$'s	$\frac{1}{8}$'s	$\frac{1}{16}$'s	$\frac{1}{32}$'s	$\frac{1}{64}$'s	...
Value of Position ⟶	2^{-1}	2^{-2}	2^{-3}	2^{-4}	2^{-5}	2^{-6}	...
How Many of Each ⟶	1	0	1	1	0	1	

A binary fraction into decimal

FIGURE 14.2

Thus the binary number 110110.101101 has the decimal equivalent 54.703125.

14.3 CONVERTING DECIMAL TO BINARY

The reverse process is somewhat more complicated. We first establish the highest position value (power of 2) which should receive a 1. Since this will be the leftmost digit of the binary number, we seek the largest power of 2 which is not bigger than the number to be converted.

As an example, let it be required to convert the decimal integer 527 into binary form. The largest power of 2 which does not exceed 527 is discovered by examining successive powers of 2, from $2^0 = 1$ on up. They are, 1, 2, 4, 8, 16, 32, 64, 128, 256, 512, 1024, ... When we reach $1024 = 2^{10}$, we exceed 527, and thus the largest power of 2 contained in 527 is $2^9 = 512$. This indicates a 1 in the leftmost position, the tenth from the right. Placing a 1 there yields

$$1 . . . .$$

and it uses up 512 of the 527 units we started out with, leaving 15 units not yet used, to be arranged in the remaining nine positions to the right.

The remaining 15 units are not enough for the next position to the right, since that is reserved for $2^8 = 256$. We thus place a 0 in that position:

$$1\ 0\ \ldots\ldots$$

The same is true for 128, 64, 32 and 16—there are not enough units in 15:

$$1\ 0\ 0\ 0\ 0\ 0\ \ldots.$$

The remaining 15 units, however, are enough for $2^3 = 8$ for which the next position is reserved:

$$1\ 0\ 0\ 0\ 0\ 0\ 1\ \ldots$$

This leaves $7 = 4 + 2 + 1$, and thus we complete the binary integer:

$$1\ 0\ 0\ 0\ 0\ 0\ 1\ 1\ 1\ 1.$$

$$
\begin{array}{rl}
527 & \\
-\ 512 & \quad 1 \\
\hline
15 & \\
-\ 256 & \\
+\ 256 & \quad 0 \\
\hline
15 & \\
-\ 128 & \\
+\ 128 & \quad 0 \\
\hline
15 & \\
-\ \ 64 & \\
+\ \ 64 & \quad 0 \\
\hline
15 & \\
-\ \ 32 & \\
+\ \ 32 & \quad 0 \\
\hline
15 & \\
-\ \ 16 & \\
+\ \ 16 & \quad 0 \\
\hline
15 & \\
-\ \ \ 8 & \quad 1 \\
\hline
7 & \\
-\ \ \ 4 & \quad 1 \\
\hline
3 & \\
-\ \ \ 2 & \quad 1 \\
\hline
1 & \\
-\ \ \ 1 & \quad 1 \\
\hline
0 & \\
\end{array}
$$

Converting 527 to binary

FIGURE 14.3

14.2 Algorithm (Decimal to Binary):   Given a positive decimal integer N (an integer in decimal notation), to obtain its binary equivalent B.

Step 1.   Initialize n to N and k to 0; i.e. $n \leftarrow N$ and $k \leftarrow 0$. (After each step, the amount left from N to be converted into binary is the value of n.) (Do step 2.)

Step 2.   If $2^{k+1} > N$, do Step 4. (Otherwise, do Step 3.)

Step 3.   Increment k by 1 (i.e., $k \leftarrow k + 1$) and do Step 2.

Step 4.   Place a 1 in the $k + $ 1st position (from the right) of B. (Do Step 5.)

Step 5.   Subtract 2^k from n; i.e., $n \leftarrow n - 2^k$. (Do Step 6.)

Step 6.   If $k = 0$, stop. Otherwise, decrement k by 1 (i.e., $k \leftarrow k - 1$) (and do Step 7).

Step 7.   If $n < 2^k$, place a 0 in the $k + $ 1st position (from the right) of B and do Step 6. Otherwise, do Step 4. ∎

Converting a decimal fraction to binary form may get quite involved, as illustrated in Figure 14.4. The decimal fraction .7201 is to be converted to a binary fraction, and thus the number of 1/2's must be found. The successful subtraction of .5 = 1/2 leaves .2201, in which we now try to fit .25 = 1/4, and so on. When the attempt is successful, a 1 is placed in the next position to the right, and when the attempt fails, 0 is

---

```
        .7201
     −  .5           1
        .2201
     −  .25
     +  .25          0
        .2201
     −  .125         1
        .0951
     −  .0625        1
        .0326
     −  .03125       1
        .00135
     −  .015625
     +  .015625      0
        .00135
     −  .0078125
     +  .0078125     0
        .00135
```

Converting .7201 to binary

FIGURE 14.4

placed in that position. As may be obvious from Figure 14.4, this process may never end. Just the same, the process is useful for approximating the given decimal fraction by a binary one.

14.4 BINARY ADDITION

The basic techniques of arithmetic with binary numbers are the familiar ones from base 10 arithmetic: carrying in addition, borrowing in subtraction, and so forth. Although a complete analysis is proper for a thorough study of the subject, illustrations of basic techniques alone should suffice for our purpose.

In adding two binary digits, as long as not both are 1, the result has only one digit. However, $2_{decimal} = 10_{binary}$ and thus when both addends are 1, we have

$$
\begin{array}{r}
1 \\
+ \quad 1 \\
\hline
10.
\end{array}
$$

The recorded digit is 0, and the 1 is "carried" to the next position on the left just as in ordinary addition base 10 (where the carried 1 stood for "ten" instead of "two").

We now demonstrate addition of two binary integers (addition of more than two should present no conceptual difficulty at this stage).

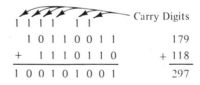

$$
\begin{array}{rr}
1\ 1\ 1\ 1 \quad 1\ 1 \qquad \text{Carry Digits} \\[2pt]
1\ 0\ 1\ 1\ 0\ 0\ 1\ 1 & 179 \\
+\quad 1\ 1\ 1\ 0\ 1\ 1\ 0 & +\ 118 \\
\hline
1\ 0\ 0\ 1\ 0\ 1\ 0\ 0\ 1 & 297
\end{array}
$$

Check: $256 + 32 + 8 + 1 = 297$

14.5 BINARY SUBTRACTION

The only comment which may be needed concerning binary subtraction is that "borrowing" a 1 from the position on the left adds a 2 to the present position. We illustrate with the reverse of the previous example:

$$1 \quad 1 \quad 2 \qquad 1 \qquad \text{Borrowing}$$

$$0 \; \cancel{2} \; \cancel{2} \; \cancel{0} \; 2 \; 0 \; \cancel{2} \; 2$$

$$\cancel{1} \; \cancel{0} \; \cancel{0} \; \cancel{1} \; \cancel{0} \; \cancel{1} \; \cancel{0} \; \cancel{0} \; \cancel{1} \qquad\qquad 297$$

$$- \qquad \underline{1 \; 1 \; 1 \; 0 \; 1 \; 1 \; 0} \qquad \underline{-\;118}$$

$$ \qquad 1 \; 0 \; 1 \; 1 \; 0 \; 0 \; 1 \; 1 \qquad\qquad 179$$

$$128 + 32 + 16 + 2 + 1 = 179$$

14.6 BINARY MULTIPLICATION

The procedure in binary multiplication is identical to that in decimal multiplication. The added convenience in the binary case is that it consists entirely of *shifting, copying,* and of course, adding: shifting one position to the left when the multiplier digit is 0 and copying when the multiplier digit is 1, as the example shows.

$$
\begin{array}{r}
1\,1\,0\,1\,1 \\
\times\,1\,1\,0\,1\,0 \\
\hline
1\,1\,0\,1\,1\,0 \\
1\,1\,0\,1\,1\,0 \\
1\,1\,0\,1\,1 \\
\hline
1\,0\,1\,0\,1\,1\,1\,1\,1\,0
\end{array}
\qquad
\begin{array}{r}
27 \\
\times\,26 \\
\hline
702
\end{array}
$$

$$512 + 128 + 32 + 16 + 8 + 4 + 2 = 702.$$

14.7 BINARY DIVISION

Binary ("long") division uses the same principles as decimal division does. Again there is an added convenience in binary division, since finding "where to start" consists of simple comparison of the divisor with the left part of the dividend. (If the first four digits form a number that is too small, use the first five digits.)

$$8 + 4 + \tfrac{1}{2} + \tfrac{1}{8} = 12\tfrac{5}{8}$$

$$
\begin{array}{r}
1\,1\,0\,0\,.\,1\,0\,1 \\
1\,1\,0\,1\,\overline{\big|\,1\,0\,1\,0\,0\,1\,0\,0\,.\,0\,0\,1} \\
\underline{1\,1\,0\,1} \\
1\,1\,1\,1 \\
\underline{1\,1\,0\,1} \\
1\,0\,0\,0\,0 \\
\underline{1\,1\,0\,1} \\
1\,1\,0\,1 \\
\underline{1\,1\,0\,1}
\end{array}
\qquad
\begin{array}{r}
12.625 \\
13\,\overline{\big|\,164.125}
\end{array}
$$

The nonending binary expansion of a "fraction" is handled in the same manner as in the decimal system. We adopt the convention that the repeating pattern, the period, in the infinitely repeating expansion may be put in parentheses followed by three dots. Thus, .00(101)... is a more precise expression for .00101101101... and a more convenient one than .00$\overline{101}$... .

To obtain the binary expansion of 5/28, we divide $5_{\text{decimal}} = 101_{\text{binary}}$ by $28_{\text{decimal}} = 11100_{\text{binary}}$, to obtain

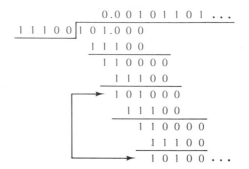

At this point, if not earlier, the remainders are seen to be repeating and therefore the digits of the quotient will also repeat. Hence, $(5/28)_{\text{decimal}} = .00(101)..._{\text{binary}}$.

14.8 OCTAL ARITHMETIC

Octal (base 8) arithmetic is fairly often the internal operation mode of computers. It therefore seems particularly suitable for illustrating arithmetic in a base higher than 2 but still different from 10.

The principles established for binary arithmetic hold for octal arithmetic as well. However, the special convenience of binary arithmetic—"if it's there at all, there is only one of it"—is no longer available.

The digits of octal arithmetic are

$$0, 1, 2, 3, 4, 5, 6, 7$$

and the value of the positions in an octal number are powers of 8. Thus, from the right, the values of the positions of an octal integer are 1; 8; 64; 512; 4,096; 32,768; 262,144 etc., as shown in Figure 14.5.

	262,144's	32,768's	4,096's	512's	64's	8's	1's	
. . .	8^6	8^5	8^4	8^3	8^2	8^1	8^0	← Value of Position
. . .	7	0	5	4	2	3	2	← How Many of Each

The octal integer $7{,}054{,}232_8$

FIGURE 14.5

As an example, the value of the octal integer $7{,}054{,}232_8$ is computed as

$$
\begin{aligned}
&7(262{,}144) + 5(4{,}096) + 4(512) + 2(64) + 3(8) + 2(1)\\
&= 1{,}835{,}008 + 20{,}480 + 2{,}048 + 128 + 24 + 2\\
&= 1{,}857{,}690.
\end{aligned}
$$

The process of converting the decimal integer $D = 1{,}857{,}690$ to an octal one, E, begins with fitting successively larger powers of 8 into the given decimal integer D until it is exceeded:

$$8^6 = 262{,}144 \leqq D \text{ but } 8^7 = 2{,}097{,}152 > D.$$

$$
\begin{array}{rll}
1{,}857{,}690 & & \\
\underline{-\ 1{,}835{,}008} & = 7(8^6) & 7\\
22{,}682 & & \\
-\ 32{,}768 & & \\
\underline{+\ 32{,}768} & & 0\\
22{,}682 & & \\
\underline{-\ 20{,}480} & = 5(8^4) & 5\\
2{,}202 & & \\
\underline{-\ 2{,}048} & = 4(8^3) & 4\\
154 & & \\
\underline{-\ 128} & = 2(8^2) & 2\\
26 & & \\
\underline{-\ 24} & = 3(8^1) & 3\\
2 & & \\
\underline{-\ 2} & = 2(8^0) & 2\\
0 & &
\end{array}
$$

Converting decimal to octal

FIGURE 14.6

Thus, the leftmost digit of E will indicate how many 8^6's (262,144's) can be fitted into $D = 1,857,690$. Since $7(8^6) \leq D$ is all that would fit, and since 7 is the largest octal digit, 7 must be the leftmost digit of E. This leaves 22,682 units from D, and that is not enough for even one $8^5 = 32,768$; thus the next digit of E is 0.

Since $5(8^4) = 20,480 \leq 22,682$ but $6(8^4) = 24,576 > 22,682$, there is room for only five of the numbers of next size, 4,096; thus, the next digit of E is 5.

The rest of E is computed in the same manner, as illustrated in Figure 14.6: $7,054,232_8$.

The arithmetic operations in octal arithmetic obey the principles established for the corresponding operations in binary arithmetic. (In the following, the subscripts indicate the number base.)

Addition (Carrying):

$$
\begin{array}{r}
\overset{1}{} \;\; \overset{1}{} \;\; \overset{1}{}\overset{1}{} \quad \text{Carry Digits}\\
7\;0\;5\;4\;2\;3\;2_8\\
+\quad 2\;1\;6\;0\;5\;6\;1_8\\
\hline
1\;1\;2\;3\;5\;0\;1\;3_8
\end{array}
\qquad
\begin{array}{r}
1{,}857{,}690_{10}\\
+\quad 582{,}001_{10}\\
\hline
2{,}439{,}691_{10}
\end{array}
$$

Subtraction (Borrowing):

$$
\begin{array}{r}
7 \quad\quad \text{Borrowing}\\
0\;9\;1\;11\;4\;\cancel{8}\;9\\
\cancel{1}\;\cancel{1}\;\cancel{2}\;\cancel{3}\;\cancel{5}\;\cancel{0}\;\cancel{1}\;3_8\\
-\quad 2\;1\;6\;0\;5\;6\;1_8\\
\hline
7\;0\;5\;4\;2\;3\;2_8
\end{array}
\qquad
\begin{array}{r}
2{,}439{,}691_{10}\\
-\quad 582{,}001_{10}\\
\hline
1{,}857{,}690_{10}
\end{array}
$$

Single-digit multiplication *(converting and carrying):* Multiple-digit multiplication is a very simple variation on single-digit multiplication, and thus we explore the latter first.

In step a (see page 261), the product $3 \cdot 4 = 12$ has more than the largest single octal digit 7. Separate the 12 into $8 + 4$; the 8 will be counted in step b as 1 in the next position (carry) and the 4 is recorded right away.

Step b represents no difficulty.

In Step c, the product 18 is $2 \cdot 8 + 2$. The $2 \cdot 8 = 16$ will be counted in step d as 2 of the next higher position (a 2-carry) and the remaining 2 is recorded now.

In step d, the 2-carry is added to the product $3 \cdot 3 = 9$ and the resulting 11 is regarded as $1 \cdot 8 + 3$. The 3 is recorded immediately and the 8 is recorded as a 1-carry in the next higher position.

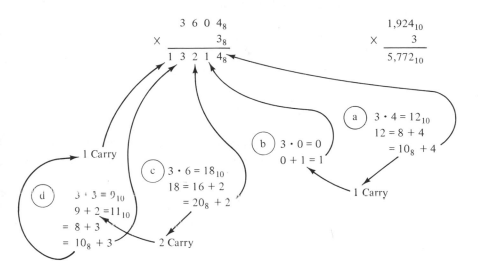

Multiple-digit multiplication (more of the above): This example consists of three single-digit multiplications and an addition.

$$
\begin{array}{r}
3\ 6\ 0\ 4_8 \\
\times \quad 2\ 6\ 3_8 \\
\hline
1\ 3\ 2\ 1\ 4 \\
2\ 6\ 4\ 3\ 0 \\
7\ 4\ 1\ 0 \\
\hline
1\ 2\ 4\ 0\ 5\ 1\ 4_8
\end{array}
\qquad
\begin{array}{r}
1,924_{10} \\
\times \quad 179_{10} \\
\hline
344,396_{10}
\end{array}
$$

Division (fitting, subtracting and borrowing):

$$
2\,6\,3_8 \overline{)\,1\,2\,4\,0\,5\,1\,4_8}^{\;3}
$$

The first three digits form a number that is too small for the divisor, but the first four digits are sufficient. The first four octal products of 263_8 are $1 \cdot (263_8) = 263_8$, $2 \cdot (263_8) = 546_8$, $3 \cdot (263_8) = 1031_8$ and $4 \cdot (263_8) = 1314_8$. The last of those is too big to fit in 1240_8, and thus the first digit of the quotient is 3, and it accounts for 1031_8 units. These 1031_8 units must be subtracted from 1240_8.

The remainder 207_8 is indeed smaller than the divisor 263_8, and so we "take down" the next digit 5 and fit as many 263_8's as 2075_8 will accomodate.

$$
\begin{array}{r}
3 \\
2\,6\,3_8 \,\overline{\smash{)}\,1\,2\,4\,0\,5\,1\,4_8} \\
\underline{1\,0\,3\,1 } \\
2\,0\,7\,5
\end{array}
$$

We continue the succession of products of 263_8: $5 \cdot (263_8)$ $= 1577_8$, $6 \cdot (263_8) = 2062_8$, $7 \cdot (263_8) = 2345_8$. The last is too big to fit into 2075_8 and thus the next digit of the quotient is 6. After subtracting $6 \cdot (163_8) = 2062_8$ from 2075_8, we have 13_8 left.

$$
\begin{array}{r}
3\,6 \\
2\,6\,3_8 \,\overline{\smash{)}\,1\,2\,4\,0\,5\,1\,4_8} \\
\underline{1\,0\,3\,1 } \\
2\,0\,7\,5 \\
\underline{2\,0\,6\,2 } \\
1\,3\,1
\end{array}
$$

Even when the next digit 1 is "taken down," there is no room in 131_8 for 263_8, and hence the next digit of the quotient is 0. After "taking down" the last digit 4, the remaining part of the dividend, 1314_8, is precisely $4 \cdot (263_8)$. The complete division then is

$$
\begin{array}{r}
3\,6\,0\,4_8 \\
2\,6\,3_8 \,\overline{\smash{)}\,1\,2\,4\,0\,5\,1\,4_8} \\
\underline{1\,0\,3\,1 } \\
2\,0\,7\,5 \\
\underline{2\,0\,6\,2 } \\
1\,3\,1\,4 \\
\underline{1\,3\,1\,4} \\
0
\end{array}
$$

14.9 ARITHMETIC IN OTHER BASES

The illustrations of arithmetic in the bases 2 and 8, in addition to the reader's knowledge of base-10 arithmetic, should be sufficient to handle arithmetic in any base. Two remarks are in order, though.

First, when the base is larger than 10, the values 10, 11, 12 ... are single digits all the way up to 1 less than the base. There are no familiar one-digit symbols for these. In base 12 (duodecimal) arithmetic, the custom is to use t for 10 ($t$en) and e for 11 ($e$leven). However, this practice fails with twelve, whose first letter t is the same as that of ten. The user must provide one-character symbols for the digits whose values are higher than 9.

The second remark concerns conversion of numbers between two bases, neither of which is 10. For example, converting the octal integer 263_8 to a quintal (base-5) integer. Techniques are available for direct conversion, but unless there are very many numbers to convert with the same two bases the practitioner might be better off using base-10 as a way station:

$$263_8 = 2(64) + 6(8) + 3 = 128 + 48 + 3 = 179_{10}.$$

The first five powers of 5 are $5^0 = 1, 5, 25, 125, 625$ and thus we compute

$$
\begin{array}{rl}
179_{10} & \\
-\ 125 & \quad 1 \\
\hline
54 & \\
-\ 50 & \quad 2 \\
\hline
4 & \\
-\ 5 & \\
+\ 5 & \quad 0 \\
\hline
4 & \\
-\ 4 & \quad 4 \\
\hline
0 &
\end{array}
$$

Hence, $179_{10} = 1204_5$, and thus $263_8 = 1204_5$.

15

The Positive Integers and Cardinalities

15.1 INTRODUCTION

Chapters 15 and 16 are interwoven and should not be regarded as independent, although they may be read separately without damage. Often one could find two treatments of the topics they include with the order of presentation reversed: what is a consequence in one is a premise in the other. Since in this book we are concerned less with correctly identifying the cart and the horse than with presenting the facts and the tools in a usable way, the reader who is accustomed to a different order of the topics should keep the following in mind: each of the chapters and topics of this book appears here to help some potential reader understand a part of mathematics without having to refer to another book.

15.2 PEANO'S AXIOMS

$\mathbb{P}$ We take for granted the reader's familiarity with the set $\mathbb{P}$ of **positive integers** (sometimes called the **counting numbers,** or **natural numbers**). The customary presentation of the positive integers is due to the Italian mathematician G. Peano (1858–1932) and is known as **Peano's axioms** or **Peano's postulates.** It is an elegant construction which sets the "rules of the game." It

avoids the use of an operation on the set $\mathbb{P}$ by using the
successor function s: $\mathbb{P} \rightarrow \mathbb{P}$ defined by

$$s(x) = x + 1.$$

So $s(x)$ is x's **successor,** in Peano's language.

With the aid of the successor function, Peano extracted
five properties of the positive integers from which all others
may be derived. They are summarized in 15.1 and may be
regarded, in a sense, as the definition of the set $\mathbb{P}$ of positive
integers.

15.1 Peano's axioms:

(i) $1 \in \mathbb{P}$.

(ii) If $x \in \mathbb{P}$, then $s(x) \in \mathbb{P}$. (Translated: if $x \in \mathbb{P}$ then $x + 1 \in \mathbb{P}$.)

(iii) There exists no $x \in \mathbb{P}$ so that $s(x) = 1$. (Translated: 1 is no one's successor.)

(iv) If $s(y) = s(x)$ then $y = x$. (Translated: different elements cannot have the same successor; or, $x \neq y \Rightarrow x + 1 \neq y + 1$.)

(v) Let $T \subseteq \mathbb{P}$, let $1 \in T$, and for every $x \in T$ let $s(x) \in T$; then $T = \mathbb{P}$.

The fifth axiom is called the **induction axiom.** It is a most
powerful tool and it leads to a most powerful method of proof
called the **principle of (finite, mathematical) induction,** which
we discuss later in some detail. It also leads to another
important property of positive integers, the **well-ordering principle,** which is also an important tool for proofs, discussed
later.

15.3 POSITIVE INTEGERS FROM THE REAL NUMBER SYSTEM

$\mathbb{R}$

Another presentation of the positive integers, both useful
and interesting, starts with axioms for a complete ordered field,
and the real number system $(\mathbb{R}, +, \cdot)$ is shown to be the only
one; i.e., any other complete ordered field is isomorphic to the
real number system. The **axioms of order** identify a subset of $\mathbb{R}$
which is the set of **positive** elements.

Then an **inductive set** A is defined as a set with the two properties:

(i) The number 1 is a member of A: $1 \in A$.

(ii) For every $x \in A$, $x + 1$ is also a member of A: $x \in A \Rightarrow x + 1 \in A$.

The reader probably recognizes the set of positive integers as an inductive set, but it is not the only one. Some other examples are: the set of all real numbers; the set $\{1\} \cup \{x: x \geq 2\}$; the set of all positive integers in union with the successors of π, i.e., $\mathbb{P} \cup \{\pi + x: x \in \mathbb{P}\}$, etc.

Now, the set $\mathbb{P}$ of **positive integers** is defined as the intersection of all inductive sets of $\mathbb{R}$, i.e., an element x of $\mathbb{R}$ is in $\mathbb{P}$ iff x is a member of every inductive set.

The axiom of induction follows now as an easy consequence, as does the well ordering of the positive integers.

This development "finds" the positive integers in the larger setting of the real number system. In contrast, the Peano axioms "invent," or construct, the positive integers without the larger context.

15.4 COUNTABLE SETS

Although we later return to well ordering and induction, our purpose in the presentation thus far (in addition to the needed basis for later development) is to present a perspective of the positive integers from two points of view that would aid the reader in understanding finite and infinite cardinality.

As was shown intuitively in discussing the set $\mathcal{P}(A)$ of all subsets of A, the cardinality of a set is its *size*, the number of elements in it. In the case of *finite sets*, this interpretation is sufficient, but more is required for infinite sets.

15.2 Definition: Let A and B be sets. A 1-1 (read "**one to one**") **correspondence** on A to B is a monic and epic function on A onto B.

The phrase "1-1 correspondence **between** A and B" is common, for the obvious reasons. We can now define cardinality in terms of 1-1 correspondence.

15.3 Definition: A set S is **finite** iff $S = \phi$ or S can be put in 1-1 correspondence with a set of the form $\{x: x \in \mathbb{P}$ and $x \leq y\}$ for some $y \in \mathbb{P}$; i.e. $\{1,2,\ldots,y\}$. The **cardinality** of S then is

$$\#(S) = y.$$

A set S is **denumerable,** or **countably infinite,** iff it can be put in 1-1 correspondence with the set $\mathbb{P}$ of positive integers, i.e. if A is the range of a monic and epic function on $\mathbb{P}$. The **cardinality** of S then is

$\aleph_0$
$$\#(S) = \aleph_0$$

(pronounced "aleph null").

A set is **countable** iff it is either finite or denumerable. Otherwise the set is said to be **uncountable.**

The language is well chosen here, since any set that is finite, or which is in 1-1 corresponds with the *counting numbers* of $\mathbb{P}$, may be used for counting.

The mention of uncountable sets, which do not 1-1 correspond to $\mathbb{P}$ or to any finite set, intimates cardinalities which are "bigger" than $\aleph_0$. Indeed, we encounter them shortly.

15.5 THE PIGEONHOLE PRINCIPLE

Infinite sets, even "simple" ones such as $\mathbb{P}$, have interesting properties that are not possessed by finite sets. For example, the **pigeonhole principle,** a *useful tool in proofs* concerning finite sets, does not hold for infinite sets. It states that, if you put $n + 1$ (or more) pigeons in n pigeonholes, you must have at least two pigeons in at least one of the pigeonholes. More formally,

15.4 Theorem: There is no 1-1 correspondence between a finite set and any one of its proper subsets.

The illustration in Figure 15.1 shows two "pigeons," c and d, in one "hole," 3.

On the other hand, we are able to display several 1-1

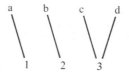

Four pigeons in three holes

FIGURE 15.1

correspondences on $\mathbb{P}$ to proper subsets and proper supersets of itself.

Mapping each positive integer to its successor results in

$$
\begin{array}{ccccccc}
1 & 2 & 3 & 4 & 5 & 6 & \cdots \\
| & | & | & | & | & | \\
1 & 2 & 3 & 4 & 5 & 6 & 7 \cdots
\end{array}
$$

(i) $\mathbb{P} \xleftrightarrow{\;1-1\;} (\mathbb{P} - \{1\})$.

Mapping each positive integer to its double, yields

$$
\begin{array}{cccccc}
1 & 2 & 3 & 4 & 5 & 6 & \cdots \\
| & | & | & | & | & | \\
1\;2 & 3\;4 & 5\;6 & 7\;8 & 9\;10 & 11\;12 & \cdots
\end{array}
$$

(ii) $\mathbb{P} \xleftrightarrow{\;1-1\;} \{2x : x \in \mathbb{P}\}$

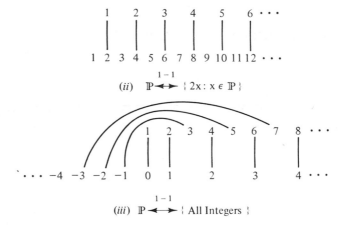

(iii) $\mathbb{P} \xleftrightarrow{\;1-1\;} \{\text{All Integers}\}$

Mapping 1 to 0 and then alternating between the next positive and negative integers establishes that the cardinality of the entire set of integers is also $\aleph_0$, as is demonstrated in (iii), above.

It is worth noting that *any scheme* of a 1-1 correspondence between two sets establishes **equality of cardinality** of the two sets.

15.6 THE RATIONAL NUMBERS HAVE CARDINALITY $\aleph_0$

The following 1-1 correspondence between $\mathbb{P}$ and the set F of positive fractions is well known. As we shall show, it may be used to prove that the entire set of rational numbers has cardinality $\aleph_0$.

We list all positive fractions whose *numerator* is 1 in column 1, those with numerator 2 in column 2, etc. At the same time, we list the fractions with *denominator* 1 (the positive integers) in row 1, those with denominator 2 in row 2, etc. In general, the fraction a/b appears in row b and column a, as follows:

$$
\begin{array}{ccccc}
\dfrac{1}{1} & \dfrac{2}{1} & \dfrac{3}{1} & \dfrac{4}{1} & \dfrac{5}{1} \quad \cdot \quad \cdot \quad \cdot \\[2ex]
\dfrac{1}{2} & \dfrac{2}{2} & \dfrac{3}{2} & \dfrac{4}{2} & \dfrac{5}{2} \quad \cdot \quad \cdot \quad \cdot \\[2ex]
\dfrac{1}{3} & \dfrac{2}{3} & \dfrac{3}{3} & \dfrac{4}{3} & \dfrac{5}{3} \quad \cdot \quad \cdot \quad \cdot \\[2ex]
\dfrac{1}{4} & \dfrac{2}{4} & \dfrac{3}{4} & \dfrac{4}{4} & \dfrac{5}{4} \quad \cdot \quad \cdot \quad \cdot \\[2ex]
\dfrac{1}{5} & \dfrac{2}{5} & \dfrac{3}{5} & \dfrac{4}{5} & \dfrac{5}{5} \quad \cdot \quad \cdot \quad \cdot \\[2ex]
\cdot & \cdot & \cdot & \cdot & \cdot \\[1ex]
\cdot & \cdot & \cdot & \cdot & \cdot \\[1ex]
\cdot & \cdot & \cdot & \cdot & \cdot
\end{array}
$$

All we need to do now is to find a way to reach each of the fractions once and only once *in turn*, so that we can associate the *first* we meet with 1, the *second* with 2, and so forth, establishing in the process a 1-1 correspondence with $\mathbb{P}$.

We do so by starting at the one fixed "finite" corner at the top left and then moving back and forth through successive diagonals, following the route indicated by the arrows:

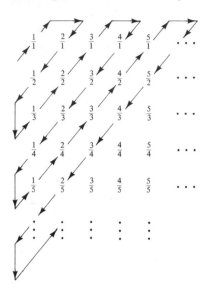

The order of encounter (and therefore the 1-1 correspondence with the positive integers) is

$$\frac{1}{1}, \frac{2}{1}, \frac{1}{2}, \frac{1}{3}, \frac{2}{2}, \frac{3}{1}, \frac{4}{1}, \frac{3}{2}, \frac{2}{3}, \frac{1}{4}, \frac{1}{5}, \frac{2}{4}, \frac{3}{3}, \frac{4}{2}, \frac{5}{1}, \ldots$$

The set F contains duplicate values, such as $1/1, 2/2, 3/3,$ ... and $1/2, 2/4, 3/6,$ ... but *it contains every positive rational number* at least once.

Still we can establish a 1-1 correspondence between it and the set of all positive rational numbers by eliminating from the list every fraction we meet whose value is already on the list. Since at each point there is only a finite number of fractions to check, the list may be established systematically, moving to the next fraction after the present one has been checked. The start of the list will then have the following appearance, with the circled fractions omitted:

$$\frac{1}{1}, \frac{2}{1}, \frac{1}{2}, \frac{1}{3}, \boxed{\frac{2}{2}}, \frac{3}{1}, \frac{4}{1}, \frac{3}{2}, \frac{2}{3}, \frac{1}{4}, \frac{1}{5}, \boxed{\frac{2}{4}}, \boxed{\frac{3}{3}}, \boxed{\frac{4}{2}}, \frac{5}{1}, \ldots$$

If we now make a similar list for the negative fractions and create a third list which starts with 0 and then alternates between the two lists, we obtain

$$0, \frac{1}{1}, -\frac{1}{1}, \frac{2}{1}, -\frac{2}{1}, \frac{1}{2}, -\frac{1}{2}, \frac{1}{3}, -\frac{1}{3}, \frac{3}{1}, -\frac{3}{1}, \cdots$$

in which every rational number must be encountered sooner or later (and only once), since it must be in the positive list or the negative list, if it is not 0.

15.7 INFINITE SETS

This property we just illustrated **characterizes** infinite sets, that is, it is possessed by all, and only by infinite sets:

15.5 Theorem: A set S is infinite iff there exists a 1-1 correspondence between S and a proper subset of itself.

We present several results on countable sets and functions on them. The proofs are omitted, and the reader's intuition is invited to substitute.

15.6 Theorem: A subset of a countable set is countable.

15.7 Theorem: If the domain of a function is countable, then the range is also countable.

15.8 Theorem: The union of a countable family of countable sets is countable.

To digest this last theorem, imagine the *sets* listed (a list is in 1-1 correspondence to the positive integers, because it has a first entry, and a second, third, etc.) in any order whatsoever, one list per row, starting on the left and proceeding to the right. The effect is identical to that of listing the positive fractions; i.e. the 5th entry of the 2nd list is in column 5 and row 2. The same scheme of traversal will work in this case as well.

We shall compare infinite cardinalities and establish which are bigger than which. In the meantime, it is not hard to see that the cardinality $\aleph_0$ of the positive integers (and any infinite subset of the positive integers, as well as the entire set of integers and the set of rational numbers) is the "smallest" infinite cardinality: if a subset of $\mathbb{P}$ is not finite, then it has a first member (the well ordering of $\mathbb{P}$) and each of its following members may be listed in their order in $\mathbb{P}$, i.e. it's an infinite *list* and therefore with cardinality $\aleph_0$.

In fact, *every infinite set has a subset whose cardinality is* $\aleph_0$.

15.9 Notation: Let A and B be sets. Then

< $$A < B$$

shall mean, "there is a monic function on A into B." We also
>   write $B > A$.

15.10 Definition: Let A and B be sets. Then, A and B are said to be **cardinally equivalent,** written

~ $$A \sim B,$$

iff there is a 1-1 correspondence between A and B (on A to B).

$\not\sim$ If A and B are not cardinally equivalent, we write $A \not\sim B$.

Cardinal equivalence is easily seen as an equivalence relation on sets. (The reader might wish to verify this.)

15.8 CANTOR'S DIAGONALIZATION ARGUMENT—THE REALS ARE BIGGER

The well known **diagonalization** argument shows that $\mathbb{P} < \mathbb{R}$ but $\mathbb{P} \not\sim \mathbb{R}$, i.e., the set of real numbers is "really bigger" than the set of positive integers. It requires several steps.

We shall demonstrate with the set $[0,1]$ of all real numbers between 0 and 1, and show that it cannot be listed *in any way*, that is, there is no way to establish a 1-1 correspondence between $[0,1]$ and $\mathbb{P}$. The set REC = $\{1/x: x \in \mathbb{P}\}$ of the reciprocals $(1/1, 1/2, 1/3, \ldots)$ of all positive integers is in a 1-1 correspondence with $\mathbb{P}$. Since REC $\subseteq [0,1]$, we shall then have $\mathbb{P} < [0,1]$ and $\mathbb{P} \not\sim [0,1]$. Thus, if $[0,1]$ is "too big," then its superset $\mathbb{R}$ certainly is.

Now, without proving the details here, we rely on the reader's acquaintance with the infinite decimal expansion of the real numbers in [0,1]: the rational ones have "terminating" expansions, i.e. with repeating zeros on the right (e.g. .3 = .3000...) or repeating nonterminating ones (e.g. 1/3 = .333...), while the irrational numbers have expansions which are nonrepeating nonterminating. (The discrepancy between equivalent expansions, for instance .2999... = .3000..., is eliminated by the fact that such cases are all rational numbers and therefore duplications may be eliminated from a list as we did above in duplicated values of the fractions.)

So, we are dealing with *all* real numbers between 0 and 1 in their decimal expansions and we want to show that it is impossible to list them *in any order*. We then suppose, by way of contradiction, that a way exists to list all these expansions. The first three real numbers on the list, no matter which ones they are, will appear as

$$.a_{11}a_{12}a_{13}a_{14}a_{15}\ldots$$

$$.a_{21}a_{22}a_{23}a_{24}a_{25}\ldots$$

$$.a_{31}a_{32}a_{33}a_{34}a_{35}\ldots$$

where a_{ij} is the j-th digit of the i-th number on the list.

Suppose that, by sheer luck, the first three numbers on the list are $\pi/10$, $1/2$, and $1/e$. Then, the beginning of the list appears as

$$.314159265\ldots$$

$$.500000000\ldots$$

$$.367879441\ldots,$$

each extending infinitely to the right, but only the 0's of $1/2$ repeating forever. Then, for example, $a_{13} = 4$, $a_{16} = 9$, $a_{21} = 5$, $a_{22} = 0$, $a_{32} = 6$, and $a_{39} = 1$.

It should be firmly fixed in the reader's mind that each row represents just *one* real number, and that the second number on the list appears in the second row.

Now, we are supposing that this list, extending infinitely down, contains *every* real number from [0,1]. Yet, we are able to produce a decimal expansion, and therefore a real number, between 0 and 1 that is *different from every number on the list.* Note that it does not matter what list is given, but once it is given we use it to produce a number that should be on the list but is not.

First, let us make up a number x from those on the list, just to illustrate a *diagonalizing* technique, by choosing the first digit a_{11} of the first number on the list as the first digit x_{11} of x, the second digit a_{22} of the second number on the list as x_{22}, and so on. Thus, $x = a_{11}a_{22}a_{33}$. . . and the n-th digit of the n-th list entry, for every n. In our example, this number will start with .307. . .

The selection was made from the diagonal, as is shown:

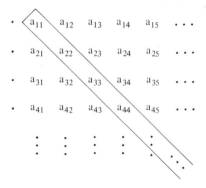

Now, let us employ the same diagonal to make up a number y in the following fashion: its first digit must be *different* from a_{11}, its second different from a_{22}, and so on, just so that its n-th digit is different from a_{nn}, for every n. We'll make it specific, to avoid arguments, by changing each digit a_{nn} into 7 unless it happens to be 7, in which case we make it 6. In our example then y will start as .776. . .

Now y is not the first real number in the list because their first digits are different: $y_{11} \neq a_{11}$. It is also not the second, since $y_{22} \neq a_{22}$. Nor is it anywhere on the list, for if it were, it would have to appear in some n-th row; but that row has a number whose n-th digit a_{nn} is different from y_{nn}.

We thus obtained a member of [0,1] which is not in the list of *all* members of [0,1]. That is a contradiction. It must be then that there is no 1-1 correspondence between $\mathbb{R}$ and $\mathbb{P}$, and that

$\mathbb{R}$ has a strictly bigger cardinality:

$$\mathbb{P} < \mathbb{R} \quad \text{and} \quad \mathbb{P} \not\sim \mathbb{R}.$$

In fact, $\mathbb{R} \sim 2^{\mathbb{P}} = \{0,1\}^{\mathbb{P}}$, the set of all functions on $\mathbb{P}$ to $\{0,1\}$. Which is to say that $\mathbb{R} \sim \mathcal{P}(\mathbb{P})$, since any function on $\mathbb{P}$ to $\{0,1\}$ is the characteristic function of a unique subset of $\mathbb{P}$.

c
$\aleph$

The **cardinality of** $\mathbb{R}$ is denoted both by c (perhaps for "continuum") and by $\aleph$. We have

$$\aleph = \#(\mathbb{R}) = \#(2^{\mathbb{P}}) = (\#2)^{\#(\mathbb{P})} = 2^{\aleph_0}.$$

Does there exist a cardinal number strictly between $\aleph_0$ and $\aleph$? That is, is there a set whose cardinality is *smaller* than that of $\mathbb{R}$ and *greater* than that of $\mathbb{P}$? The **continuum hypothesis** is the conjecture that $\aleph$ is the smallest cardinal number greater than $\aleph_0$. That is, every infinite subset of $\mathbb{R}$ has cardinality $\aleph_0$ or $\aleph$.

*15.9 HIGHER CARDINALITIES

By Cantor's theorem (15.15, below), there is an infinite succession of ever increasing infinite cardinalities, which are denoted by $\aleph_0, \aleph_1, \aleph_2, \aleph_3, \ldots$ In fact, *any set of cardinals is well ordered*. (See Sections 15.1 and 16.5.) Thus, not only is

$$\aleph_0 < \aleph_1 < \aleph_2 < \aleph_3 < \ldots$$

but $\aleph_1$ is the *immediate successor* of $\aleph_0$ and, in general, $\aleph_{i+1}$ is the immediate successor of $\aleph_i$.

The continuum hypothesis may then be phrased as the statement that

$$2^{\aleph_0} = \aleph_1.$$

Again, by Cantor's theorem (15.15), for all $i \geq 0$,

$$\aleph_{i+1} \lesssim 2^{\aleph_i}.$$

---

*This section requires somewhat greater mathematical maturity. It may also be encountered somewhat less frequently than other topics.

The **generalized continuum hypothesis** is the assertion that there exist no cardinal numbers strictly between $\aleph_i$ and $2^{\aleph_i}$ for any $i \geq 0$; that is,

$$\aleph_{i+1} = 2^{\aleph_i}.$$

This is a *hypothesis*; it is consistent with the axioms of set theory but is independent of them.

**15.10 COMBINATIONS AND TABLES OF CARDINALITIES

This concludes our discussion on cardinality. We follow it by a list of results and cardinality tables which might aid the reader who wishes to go a bit beyond this point.

15.11 Theorem: Let A and B be finite sets. Then,

$$A \cup B, A \cap B, A - B, A \times B, A^B, \text{ and } \mathcal{P}(A)$$

are finite. Furthermore,

$$|A \cup B| = |A| + |B| - |A \cap B|,$$
$$|A^B| = |A|^{|B|},$$

and

$$|\mathcal{P}(A)| = 2^{|A|}.$$

15.12 Theorem: Let A be a nonfinite subset of $\mathbb{P}$. Then $A \sim \mathbb{P}$.

15.13 Theorem (Schröder-Bernstein):   Let A and B be sets, and let $C \subseteq A$ and $D \subseteq B$ be such that $A \sim D \subseteq B$ and $B \sim C \subseteq A$. Then $A \sim B$.

15.14 Theorem: Let $A \sim C$ and $B \sim D$. Then,

$$A \times B \sim C \times D, A^B \sim C^D, \text{ and } \mathcal{P}(A) \sim \mathcal{P}(C).$$

15.15 Theorem (Cantor):   For any set A,

$$\{0,1\}^A \not\precsim A.$$

**This section requires considerable mathematical maturity. It is also quite unlikely to be encountered in the common pursuit of computer science.

In the cardinality tables below we observe the following **conventions:**

n denotes any nonzero finite cardinality,

P denotes the cardinality of a set $\sim \mathbb{P}$ (i.e. $\aleph_0$),

R denotes the cardinality of a set $\sim \mathbb{R}$ (i.e. $\aleph$), and

2^R denotes the cardinality of a set $\sim \{0,1\}^{\mathbb{R}}$.

TABLE 15.1

CARDINALITIES

(i) $A \times B$

A\B	ϕ	n	P	R
ϕ	ϕ	ϕ	ϕ	ϕ
n		n	P	R
P			P	R
R				R

(ii) $\mathcal{P}(A)$

A	$\mathcal{P}(A)$
n	n
P	R
R	$2^R \gneq R$

(iii) A^B

A\B	ϕ	n	P	R
ϕ	1	ϕ	ϕ	ϕ
n	1	n	1 or R	1 or 2^R
P	1	P	R	2^R
R	1	R	R	2^R

(iv) $A - B$

A\B	n	P	R
n	n	n	n
P	P	n or P	n or P
R	R	R	n or P or R

(v) $A \cap B$

A\B	n	P	R
n	n	n	n
P		n or P	n or P
R			n or P or R

(vi) $A \cup B$

A\B	n	P	R
n	n	P	R
P		P	R
R			R

16

Tools and Techniques
for Proving Theorems

16.1 INTRODUCTION

Of all the tools and methods of proof available in the mathematical arsenal, the ones used commonly in computer science and its applications dictated the selection in this chapter.

The chapter consists of thirteen sections which concentrate on methods of proof and concepts needed for these methods. Chief among them are the pigeonhole principle, the well-ordering principle for positive integers, existence proofs, proofs by counterexample, finite induction, proof by induction, the division theorem, direct and indirect proofs, and proofs by contradiction.

Many of the items are stated, with little or no accompanying comment, while some are stated with explanations, examples (however primitive), and points of caution. The reader who is not already well-versed in proving theorems and in understanding such proofs, may find it profitable to return to this chapter several times, bringing each time more experience and greater facility.

16.2 THE PIGEONHOLE PRINCIPLE

Theorem 15.4 gives the formal statement of this principle, whose essence is that a function on a finite set to a *smaller* finite set cannot be monic. In other words, if $k + 1$ pigeons are forced into k pigeonholes, at least one hole contains two or more pigeons.

A simple example of the pigeonhole principle is a function on a finite set, whose range is smaller than its domain.

When the domain D of f is finite and $\#(f(D)) < \#D$, then f *cannot be monic*. It is not necessary to know which two elements are mapped identically, but the principle asserts that *there exist two elements of D which are mapped to the same element of C.*

16.3 EXISTENCE PROOFS; THE ARCHIMEDEAN PROPERTY

It is common to encounter a theorem which asserts the *existence* of an entity possessing a certain property. To prove such a theorem, it is sufficient to exhibit the claimed entity, and to show that it possesses the property.

We take as an example a basic property of the positive integers: it is always possible to multiply one of them enough to make the product bigger than another positive integer, no matter which two are selected.

> **16.1 Theorem (Archimedean Property):** Let a and b be positive integers. Then there exists a positive integer n such that $na > b$.
>
> *Proof:* Let $n = b + 1$. Then, $na = (b + 1)a = a(b + 1) = ab + a > ab$. But since a is a positive integer, $a \geq 1$ and hence $ab \geq b$. We then have $na > b$. ∎

The proof consists primarily in naming the required factor n in terms of b ($n = b + 1$) and showing that it would always perform the required task.

16.4 PROOF BY COUNTEREXAMPLE; QUANTIFICATION

A proof by counterexample is very much like an existence proof, in the sense that in both cases it is sufficient to exhibit one example. (There are, of course, *analytical* proofs for existence theorems, and at times they are necessary.) The difference between the two types may be considered semantic only; it depends to an extent on the way the theorem is stated: "It is false that every *A* has property *p*" may be proved by exhibiting an *A* (whatever *A* happens to represent) which does not have the property.

There are some who would object to calling this *example* a *counter*example, and would reserve the latter term for disproofs, i.e. refutations of statements. Thus, the statement, "every *A* has property *p*" may be disproved by exhibiting a counterexample, i.e. an *A* which does not have the property. As may be seen from the laws of logic, in Section 16.7 below, the same statement is proved in both cases, and therefore we do not distinguish between proof by example and (dis-)proof by counterexample.

As an illustration, consider the statement:

"The identity element *e* of a subgroup *G* of a monoid *M* need not be the identity element of *M*."

To prove the statement, it is sufficient to display a monoid *M* and its subgroup *G*, as in Figure 16.1, with the (unique) monoid identity *i* and the group identity *b* distinct. (It is still necessary to convince the reader of the proof that *M* is monoid, *G* a group, etc., but we shall not do so here.)

	i	a	b	c	d
i	i	a	b	c	d
a	a	b	a	d	c
b	b	a	b	c	d
c	c	a	b	c	d
d	d	b	a	d	c

	a	b
a	b	a
b	a	b

(*i*) M　　　　　　(*ii*) G

A monoid M and its subgroup G

FIGURE 16.1

Definitions and statements of theorems and their proofs often include phrases such as "for all $x \in S$" and "there exists $x \in S$." These are **quantified** statements, telling for how many (what *quantity*) of the elements of *S* the statement is claimed to be true. "For all ..." is called a **universal quantifier** (since the statement is claimed *universally* for all members of S); "there exists ..." is called an **existential quantifier** (for obvious reasons).

$\forall$

The symbol for the *universal quantifier* is "$\forall$." It is used in contexts such as

$$\forall x \in S, f(x) \in T,$$

and is read, **"for all** x in S," **"for each** x in S," **"for every** x in S," and the like.

$\exists$

The symbol for the existential quantifier is "$\exists$." It is used in contexts such as

$$\exists x \in S, \text{ such that } f(x) \in T,$$

and is read **"there exists** an x in S," **"for some** x in S," and the like.

A quantified statement is still a proposition (see Section 16.7, below) which may be true or false. However, a universally quantified statement is true iff the part which follows the quantifier *actually is true for every member of the set.* In the above example, the statement

$$\forall x \in S, f(x) \in T$$

is true if $f(x) \in T$ for every $x \in S$; the entire statement is false if "$f(x) \in T$" is false even for one single $x \in S$.

To prove such a universally quantified statement false, it is sufficient to produce a *counterexample*—a single member x in S for which $f(x) \notin T$, in the example.

To prove a statement with an existential quantifier, it is sufficient to exhibit an element ($x \in S$, in our example) which satisfies the condition that follows the quantifier ($f(x) \in T$, in our example); in essence then it is an *existence proof.*

The proof that a universally quantified statement is true is usually analytic since there are too many values for the quantified variable to exhibit, often infinitely many. Likewise, the denial of an existential statement is likely to require analytical proof, as "an element does not exist such that ..." still means "for every element, it is false that ..." and we are back to verifying a universal statement.

16.5 THE WELL-ORDERING PRINCIPLE FOR POSITIVE INTEGERS

This well-ordering principle has been mentioned in Chapter 15 as a consequence of Peano's induction axiom in one treatment of the positive integers, and as a consequence of the axioms of order in another. Its statement is simple and intuitively plausible:

Every nonempty set of positive integers has a least member.

This principle is a very useful tool. We illustrate its use first in proving the division theorem, and later in proving the principle of induction.

16.6 DIRECT PROOF; THE DIVISION ALGORITHM

We later discuss indirect proofs, at which time the difference between direct and indirect proofs may be seen more vividly. In essence, in a direct proof we start with the premise(s) of the theorem and proceed to deduce the consequence, while in an indirect proof we start with the negation of the consequence and arrive at the negation of the premise, i.e. we prove the contrapositive, as explained below.

The **division algorithm** (for nonnegative integers) is actually the "long division" process, which we use to obtain the quotient and remainder. It is based on the **division theorem,** also called the **fundamental theorem of Euclid.** It states that, for each two positive integers a and b, there exist both a unique quotient q and a remainder r, each either 0 or positive.

We shall use the division theorem to illustrate both the use of the well-ordering principle and the direct proof method.

To aid the inexperienced reader, we will present the proof on the left half of the page, while on the right we will illustrate the proof with an example.

16.2 The Division Theorem (Euclid):   Let a and b be positive integers. Then there exist unique non-negative integers q (the quotient) and r (the remainder), such that $r < a$ and $b = qa + r$.

16.2 The Division Theorem (Euclid)—Continued

Proof: (We prove the existence half, and leave the uniqueness for a possible exercise for the reader who managed to reach the end of this section.)

Let a and b be given. Then, by the Archimedean property (Theorem 16.1), there is a positive integer p such that $pa > b$. Hence, the set S of positive integers p such that $pa > b$ is nonempty, and therefore it must have a least member, by the well-ordering principle. Call this least member m.　Now we let $q = m - 1$. Since m is a *positive* integer, $q = m - 1 \geqq 0$. We separate into two cases:	Suppose $a = 3$ and $b = 13$, so we seek $13 = q \cdot 3 + r$. (Clearly $q = 4$ and $r = 1$, and $13 = 4 \cdot 3 + 1$.) E.g. $p = 6$, so $pa = 6 \cdot 3 = 18 > 13 = b$. All multiples of 3 are 3, 6, 9, 12, 15, 18, ... $S = \{5,6,7,...\}$. $m = 5$, since $5 \cdot 3 = 15 > 13$ but $4 \cdot 3 = 12 < 13$. Note that the desired quotient 4 is $m - 1 = 5 - 1$.
Case 1: If $q > 0$, then $qa \leqq b$. The reason is that m is the smallest positive integer with $ma > b$, and $q = m - 1 < m$. That leaves two possibilities: $qa = b$ and $qa < b$. If $qa = b$, we let $r = 0$, and if $qa < b$ we let $r = b - qa$. The required conditions are met.	Recall $a = 3$, $b = 13$, $q = 4$, and $r = 1$. Also recall $m = 5$. $ma = 15 > 13$ but $qa = 12 < 13$. $qa < b$ since $12 < 13$. Let $r = b - qa = 13 - 12 = 1$.
Case 2. If $q = 0$, then $m = 1$ and $a > b$. So take $r = b$ and the conditions are met. ∎	New example: $a = 6$, and $b = 2$. Then $2 = q \cdot 6 + r = 0 \cdot 6 + 2$, forcing $q = 0$ and $r = 2$.

16.7 SOME LAWS OF LOGIC; PROOF BY TRUTH TABLES

Before we proceed with other methods of proof, we need to acquire some logical tools.

A **proposition** is a declarative statement which may be true or false.

~, ¬　　A **negation** of a proposition P, written $\sim P$, or $\neg P$, and pronounced "not P" is a declaration that P is false; so $\sim P$ is true iff P is false.

A **conjunction** of two propositions P and Q is a proposition of the form P & Q, read "P and Q," and is true iff *both P and Q* are true.

A **disjunction** of P and Q is a proposition of the form $P \vee Q$, read "P or Q" and is false iff both P and Q are false.

A **tautology** is a proposition which is always true regardless of the truth of its component propositions.

A **contradiction** or **inconsistency** is a proposition which is always false.

An **implication**[1] is a proposition of the form $P \Rightarrow Q$, read "P implies Q" or "if P then Q." It is false iff P is true and Q is false. (Note: knowing that $P \Rightarrow Q$ is true, we still know nothing about the truth of either P or Q.)

The **converse** of the implication $P \Rightarrow Q$ is $Q \Rightarrow P$. It is false iff Q is true and P is false. (*Note that the truth of an implication does not guarantee the truth of its converse.*)

$\equiv$ The **bi-implication**, or **equivalence**, of the propositions P and Q is the proposition $P \Longleftrightarrow Q$ or $P \equiv Q$, pronounced "P if and only if Q" or "P is equivalent to Q." P and Q always have the same truth values (either both true or both false).

The **contrapositive** of the implication $P \Rightarrow Q$ is $\sim Q \Rightarrow \sim P$. It is true iff $P \Rightarrow Q$ is true.

The terminology was presented in a terse form as it is intended only to remind the reader of the basic relationships. Truth tables for these logical connectives, which are largely self explanatory, are given in Table 16.1. (The reader who desires

TABLE 16.1

TRUTH TABLES FOR LOGICAL CONNECTIVES

P	Q	$\sim P$	P&Q	$P \vee Q$	tauto-logy	contra-diction	$P \Rightarrow Q$	$Q \Rightarrow P$	$Q \Longleftrightarrow P$	$\sim Q \Rightarrow \sim P$
T	T	F	T	T	T	F	T	T	T	T
T	F	F	F	T	T	F	F	T	F	F
F	T	T	F	T	T	F	T	F	F	T
F	F	T	F	F	T	F	T	T	T	T

---

[1]The term *implication* ($\Rightarrow$) is used here to express both the relation between antecedent and consequent of a conditional sentence, as well as the relation between the premises and conclusion of a valid argument. The reader who is sensitive to the difference should find the distinction easy from the context. A careful distinction between the two uses would result in an exposition too long and complicated for this presentation and even then it may not improve the reader's skill.

further treatment is referred to the bibliography.) We let P and Q be two arbitrary propositions and assign to them all the possible combinations of truth values: T = true and F = false.

Some of the basic laws of logic are given below. They are, of course, tautologies, and are useful in proofs of theorems when they are carefully interpreted.

(i) $\sim(\sim P) \equiv P$ (**Double negation**)

(ii) $\sim(P \ \& \ \sim P)$ (**Denial of contradiction**) "A proposition and its negation cannot both be true," or "A proposition cannot be both true and false."

(iii) $P \vee \sim P$ (**Excluded middle**) "A proposition is either true or false—there is nothing in between," or "if a proposition is false, then its negation is true."

(iv$_a$) $\sim(P \vee Q) \equiv (\sim P) \ \& \ (\sim Q)$ ⎫
(iv$_b$) $\sim(P \ \& \ Q) \equiv (\sim P) \vee (\sim Q)$ ⎬ (**DeMorgan's Laws**)

(v) $(P \Rightarrow Q) \equiv (Q \vee \sim P)$

(vi) $(P \Rightarrow Q) \equiv ((\sim Q) \Rightarrow (\sim P))$ (**Contrapositive**)

(vii) $(P \Leftrightarrow Q) \equiv ((P \Rightarrow Q) \ \& \ (Q \Rightarrow P))$

(viii) $(P \Rightarrow Q \ \& \ Q \Rightarrow R) \Rightarrow (P \Rightarrow R)$ (**Transitivity of implication**)

(ix) $\sim(P \Rightarrow Q) \equiv P \ \& \ (\sim Q)$ (**Negation of implication**)

Note that the negation $\sim(P \Rightarrow Q)$ of the implication $P \Rightarrow Q$ may also be written as $P \not\Rightarrow Q$. Its truth table is contrasted with those of $P \Rightarrow Q$ and $Q \Rightarrow P$:

P	Q	$P \Rightarrow Q$	$P \not\Rightarrow Q$	$Q \Rightarrow P$
T	T	T	F	T
T	F	F	T	T
F	T	T	F	F
F	F	T	F	T

The reader may have already identified the translation of symbols and concepts which make Table 4.1 of the laws of operating with sets applicable to the **calculus of propositions.** If not, it is an easy and interesting exercise.

A point of caution: Many proofs use a *succession of implications,* usually based on the *transitivity of implication* (viii), without pointing out the implied (and usually missing) parentheses. For instance, $P \Rightarrow Q \Rightarrow R$ often means

$[(P \Rightarrow Q) \,\&\, (Q \Rightarrow R)] \Rightarrow (P \Rightarrow R)$, while it could be interpreted as $(P \Rightarrow Q) \Rightarrow R$ or as $P \Rightarrow (Q \Rightarrow R)$, all of which are different propositions with different truth tables, shown in Table 10.5. It should be noted, in addition to the fact that "$\Rightarrow$" *is not associative,* that there is also possible confusion with the transitivity statement, i.e. the *conjunction* of two implications *implying* a third.

The three propositions are evaluated in Table 16.2.

(The final column of truth values for each proposition is indicated by an arrow. It is arrived at step by step from the columns of values of the variables P, Q, and R on the left of the table. The most "internal" connectives are evaluated first, and the resulting columns of truth values are entered under the appropriate connective; they give the values of the entire proposition spanned by these connectives. In this manner, larger and larger component propositions are evaluated, until the values of the entire proposition appear in the column under the main connective.)

TABLE 16.2

NEEDED CARE IN CHAINED IMPLICATIONS

P	Q	R	$(P \Rightarrow Q) \Rightarrow R$		$P \Rightarrow (Q \Rightarrow R)$		$((P \Rightarrow Q) \,\&\, (Q \Rightarrow R)) \Rightarrow (P \Rightarrow R)$				
T	T	T	T	T	T	T	T	T	T	T	T
T	T	F	T	F	F	F	T	F	F	T	F
T	F	T	F	T	T	T	F	F	T	T	T
T	F	F	F	T	T	T	F	F	T	T	F
F	T	T	T	T	T	T	T	T	T	T	T
F	T	F	T	F	T	F	T	F	F	T	T
F	F	T	T	T	T	T	T	T	T	T	T
F	F	F	T	F	T	T	T	T	T	T	T

The moral of the table is clear: the three propositions are not the same; therefore there is room for error, and caution should be exercised.

16.8 NEGATION OF PROPOSITIONS; PROVING IMPLICATIONS

The negation of a proposition is a starting point in proofs by contradiction, and in indirect proofs, which exploit the contrapositive and its equivalence to the original implication. For that reason, we elaborate on forming the negation of each of several basic propositions, already presented. We also show how to form the negation of a *quantified* statement.

In general, the phrase "it is false that" preceding any proposition yields the negation of the proposition. The negation of a *negation* ((i) above) requires no elaboration, as $\sim(\sim P) = P$ says it all.

The negation of a *conjunction* is given symbolically in (iv_b) as

$$\sim(P \ \& \ Q) \iff (\sim P) \lor (\sim Q).$$

In words, "it is false that both P and Q hold" is equivalent to "at least one of P and Q is false."

The negation of the *disjunction* (iv_a) is the other DeMorgan Law:

$$\sim(P \lor Q) \iff (\sim P) \ \& \ (\sim Q),$$

or, "it is false that either P or Q is true" is equivalent to "both P and Q are false."

The negation of a *tautology* is a contradiction and the negation of a *contradiction* is a tautology.

The negation of the *implication* ((ix), above) should receive special attention. First, the implication itself $P \Rightarrow Q$ is a proposition which may be true or false, depending on the *joint* values of P and Q. By the truth table (Table 16.1), the implication $P \Rightarrow Q$ is false only when P is true and Q is false at the same time.

For example, the pronouncement, "If I get an A on the final exam, I shall get an A for the course" makes a prediction (which may be right or wrong) for only *one* eventuality—that I get an A on the final exam. In that case, the statement turns out to be right if I also get an A for the course, and wrong if I do not. No promise is made concerning what happens if I do not get an A on the final exam. Thus, the pronouncement *cannot be wrong* if the **antecedent** or **premise** (P) is false, i.e. if I do not get an A on the final exam.

As it turns out, there is no way for the pronouncement to be wrong when the **consequent** or **conclusion** (Q) is true. The only time the pronouncement is false is when the promise is not kept, that is, the premise P is true but the consequent Q is false, in spite of the promise "$P \Rightarrow Q$."

Thus, there are several standard ways to prove an implication, among which are:

(a) assume P true and prove that Q must be true,

(b) assume Q false and prove that P must be false (the *contrapositive*),

(c) show that P is always false (if such is the case), or that Q is always true (if such is the case),

(d) show that P true and Q false cause a contradiction (i.e., $\sim(P \& (\sim Q))$).

The preceding remarks are easily translated for the *negation of implication*, $P \not\Rightarrow Q$, or $\sim(P \Rightarrow Q)$, read "P does not imply Q" or "it is false that P implies Q." For example, to prove that "P does not imply Q" we might show that P is true, yet Q is false.

The negation of the *converse* $Q \Rightarrow P$ of the implication $P \Rightarrow Q$ is $Q \not\Rightarrow P$ or $\sim(Q \Rightarrow P)$ and is similar to the cases of the preceding paragraphs.

The *bi-implication* (or **biconditional**) $P \Longleftrightarrow Q$ is read "P if and only if Q." It serves in very similar, even identical, roles as the *equivalence $P \equiv Q$*, read "P is equivalent to Q." Their negation is called the **exclusive or** and is denoted by $P \oplus Q$, or $P \not\equiv Q$, or $P \not\Longleftrightarrow Q$. In words, "$P$ is not equivalent to Q" or "either P or Q *but not both*," whence comes the word *exclusive*.

It is then also the case that the negation of the *exclusive or* is the equivalence, as may be seen from the truth tables:

P	Q	$P \Longleftrightarrow Q$	$P \oplus Q$
T	T	T	F
T	F	F	T
F	T	F	T
F	F	T	F

The contrapositive of the implication $P \Rightarrow Q$ is $(\sim Q) \Rightarrow (\sim P)$ and is equivalent to the implication (by (vi), above). This equivalence suggests proving "P implies Q" by assuming that Q is false and proving that P is also false by necessity. (This device is the centerpiece of **indirect proofs.**) As a result, the negation of the *contrapositive* is the same as the negation of the implication; i.e.,

$$((\sim Q) \not\Rightarrow (\sim P)) \equiv P \not\Rightarrow Q.$$

The negation of a *universal statement* such as ($\forall s \in S$, $f(s) \in T$), or "for every $s \in S, f(s) \in T$" is aimed at the *universality* of the proposition $f(s) \in T$, not at the proposition itself. What is claimed false is not that $f(s) \in T$ is true, but that it is true *for every* $s \in S$. Hence, its negation is the existence of one member of S which violates $f(s) \in T$. In symbols,

$$\sim(\forall s \in S, f(s) \in T) \equiv (\exists s \in S \text{ such that } f(s) \notin T).$$

For example, the statement, "Every soldier in this company volunteered" has the negation, "A member of this company was drafted."

Similarly, the negation of an *existential statement,* such as ($\exists s \in S$ such that $f(s) \in T$), or, "there exists a member of S which f maps into T," is aimed not at the truth of $f(s) \in T$ but at the *existence* of an s in S for which it is true. Thus, the negation would be, "for every $s \in S, f(s) \notin T$." In symbols,

$$\sim(\exists s \in S \text{ such that } f(s) \in T) \equiv (\forall s \in S, f(s) \notin T).$$

The negation of a long and complicated proposition may be obtained simply, step by careful step, using the information in the preceding paragraphs. We illustrate with an example, which should be followed only by the most tenacious reader. (We shall omit the phrase "such that" after an existential quantifier for convenience.)

Let us obtain the negation of the proposition $\mathcal{H}$ below. (It has no intended meaning; the functions f and g are not detailed.)

$$\mathcal{H}: (\forall x \in A, \ f(x) = g(x)) \Rightarrow [(\exists y \in B) \ (\forall x \in A,$$
$$f(x) \neq g(y)) \vee (A \cap B = \varnothing)].$$

We abbreviate $\mathcal{H}$ into the form:

$$(\forall x, P(x)) \Rightarrow [\exists y, (\forall x, Q(x,y)) \vee R].$$

We proceed with forming the negation $\sim\!\mathcal{H}$ of $\mathcal{H}$, working the negation from the outside into the basic small atomic propositions.

$$\sim\!\mathcal{H} \equiv \sim\![(\forall x, P(x)) \Rightarrow [\exists y, (\forall x, Q(x,y)) \vee R]]$$

$$\equiv (\forall x, P(x)) \,\&\, \sim\![\exists y, (\forall x, Q(x,y)) \vee R]$$

$$\equiv (\forall x, P(x)) \,\&\, [\sim\!(\exists y, (\forall x, Q(x,y))) \,\&\, \sim\!R]$$

$$\equiv (\forall x, P(x)) \,\&\, [(\forall y, (\sim\!(\forall x, Q(x,y)))) \,\&\, \sim\!R]$$

$$\equiv (\forall x, P(x)) \,\&\, [(\forall y, (\exists x, \sim\!Q(x,y))) \,\&\, \sim\!R],$$

which we may now expand in terms of the original entities, to obtain:

$$\sim\!\mathcal{H}: (\forall x \in A, f(x) = g(x)) \,\&\, (\forall y \in B, \exists x \in A, f(x) = g(y)) \,\&\, (A \cap B \neq \varnothing).$$

16.9 PROOFS BY CONTRADICTION

We describe what happens in a typical proof by contradiction (*reductio ad absurdum,* in Latin) before analyzing it.

A theorem often has one or more stated premises and a conclusion implied by them; the conclusion holds when all premises hold. A proof by contradiction takes all the premises of the theorem as premises and *adds* to them as a *premise the negation of the conclusion.* It is then a matter of successive implications until a contradiction of the form $R \,\&\, \sim\!R$ is reached, where R is usually one of the original premises of the theorem, but may be any theorem (a proven, true statement).

We present an example below, but first we repeat the description with symbols. The theorem is of the form $(p_1 \,\&\, p_2 \,\&\, \ldots \,\&\, p_n) \Rightarrow c$, where each p_i is a premise and their conjunction implies the conclusion c. We know (we hope) the entire implication to be a tautology, therefore the negation of the implication must be a contradiction. The negation is

$$\sim\!((p_1 \,\&\, p_2 \,\&\, \ldots \,\&\, p_n) \Rightarrow c)$$

$$\equiv p_1 \,\&\, p_2 \,\&\, \ldots \,\&\, p_n \,\&\, \sim\!c, \text{ by (ix) above.}$$

At this point, a succession of well-chosen implications (using the transitivity of implication of (viii), above) will lead to p_i & $\sim p_i$ for one of the premises p_i, or the negation of some established theorem, either of which is a contradiction. We then have

$$(p_1 \ \& \ p_2 \ \& \ \ldots \ \& \ p_n \ \& \ \sim c) \Rightarrow \ldots \Rightarrow \text{contradiction,}$$

and thus $(p_1 \ \& \ p_2 \ \& \ \ldots \ \& \ \sim c)$ must itself be a contradiction. But that contradiction is the negation of the original implication, which therefore must be true (law of the excluded middle). In summary, we used the negation of implication (ix) by observing that $P \Rightarrow Q$ is true iff P & $(\sim Q)$ is false, and by *proving* that P & $(\sim Q)$ is false.

We now combine several of the ideas of the preceding discussion (in particular, the well-ordering principle and proof by contradiction) to prove another very powerful result, the theorem of induction.

16.10 MATHEMATICAL INDUCTION

One type of theorem, which often poses a difficulty in approach, is a statement concerning an infinite number of cases. For example, let Q be the statement:

Q: The sum of n positive real numbers is positive, for any positive integer n.

Now, Q is a statement with a universal quantifier, since it may be written as

$$\text{``} \forall n \in \mathbb{P}, R(n) \text{''}$$

where $R(n)$ is the *unquantified* statement: "the sum of n positive real numbers is positive." Thus, Q is an infinite number of statements: $R(1)$, $R(2)$, $R(3)$, and so on without end. (For example, $R(2)$ is: "the sum of *two* positive real numbers is positive.")

Clearly, we cannot prove Q by proving each $R(i)$, since we shall never finish the task. We must find a device which is *finite* itself but which proves the truth of *infinitely many statements.*

Such a device is **mathematical induction.** We might view this procedure as a type of *contagion:* first we prove that $R(1)$ is true and then prove that the truth of $R(k)$ for any positive integer k will "infect" the next integer, the successor $k + 1$, with the truth of R; then we have $R(1) \Rightarrow R(2)$, $R(2) \Rightarrow R(3)$, $R(3) \Rightarrow R(4)$, and so on forever.

If the reader is still not intuitively convinced that the device we just described guarantees the truth of $R(k)$ *for every positive integer k,* let us suppose that t is such a positive integer for which $R(t)$ is false. But there is only a finite number of positive integers smaller than t and therefore t must be reached by the "contagion process" in a finite number $(t - 1)$ of steps. Hence $R(t)$ must be true.

The preceding paragraph establishes that

$$\sim(\exists t \in \mathbb{P} \text{ such that } \sim R(t)).$$

But that is equivalent to

$$\forall t \in \mathbb{P}, \sim(\sim R(t))$$

which is equivalent to

$$\forall t \in \mathbb{P}, R(t),$$

and that is precisely the proposition Q.

We now formalize the idea by stating and proving the induction theorem and then by using it as a method of proof.

> *16.3 The Induction Theorem:* Let S be a set of positive integers with the following two properties:
>
> *(i)* $1 \in S$;
> *(ii)* $\forall k \in \mathbb{P}, (k \in S \Rightarrow k + 1 \in S)$.
>
>   Then $S = \mathbb{P}$, i.e. *every* positive integer is in S.

Proof (by contradiction):We suppose that the premise is true (i.e. that the set S satisfies (i) and (ii)), but that the conclusion is false (i.e. that $S \neq \mathbb{P}$). (The theorem is of the form of an implication (premise $\Rightarrow$ conclusion) and we have formed its negation (premise & $\sim$conclusion).)

Since $S \neq \mathbb{P}$ and $S \subseteq \mathbb{P}$ (by hypothesis, and thus in the premise), there must be at least one positive integer missing from S, i.e. $\exists n \in \mathbb{P} - S$. But that means that the set $\mathbb{P} - S$ is a *nonempty subset of the positive integers* and, by the well-ordering principle, it must have a *least member*.

Let r be the least member of $\mathbb{P} - S$. Now r must be greater than 1, because $1 \in S$, by (i), and cannot be in $\mathbb{P} - S$. Therefore $r - 1$ is still a positive integer, since it is greater than 0.

Now, $r - 1$ cannot be in $\mathbb{P} - S$ because it is smaller than the smallest member r of $\mathbb{P} - S$. Hence $r - 1 \in S$.

At this point we use property (ii) of the premise to insist that r is also a member of S, since (ii) forces "$r - 1 \in S \Rightarrow (r - 1) + 1 \in S$" and since $(r - 1) + 1 = r$.

Now we have both "$r \in S$" (which we just deduced) and "$r \notin S$" (because it is in $\mathbb{P} - S$). The form is "$P \& \sim P$," that is a proposition AND its negation must both be true, *which is a contradiction.*

Consequently, our original negation of the induction theorem is false, and hence the theorem itself is true. ∎

16.11 PROOF BY MATHEMATICAL INDUCTION

The induction theorem furnishes precisely the type of proof by "contagion" which we sought for Q in Section 16.10 above. This method of proof is variously called the **principle of induction, finite induction** (as distinguished from transfinite induction), **mathematical induction,** and just **induction.** It uses the induction theorem in a very simple but ingenious way, which we now illustrate:

We hope that the statement Q, which we want to prove, is really true; so we try to identify the set of all positive integers for which Q is *actually* true (hoping this set contains every positive integer). Let us call this set $T(Q)$, so that

$$T(Q) = \{n \in \mathbb{P}: R(n) \text{ is true}\}$$
$$= \{n \in \mathbb{P}: Q \text{ is true for } n\}.$$

The actual statement was "$n \in T(Q)$ iff the sum of n

positive real numbers is positive."

We can easily establish for any *finite* collection of positive integers that they are members of $T(Q)$. In particular

$$1 \in T(Q) \text{ and } 2 \in T(Q),$$

a fact which we use later in the proof. Now, let $k \in T(Q)$. (We may say so because we already know of at least one member of $T(Q)$.)

Then, for any k positive real numbers $x_1, x_2, \ldots, x_k$ we have that

$$\sum_{i=1}^{k} x_i = x_1 + x_2 + \ldots + x_k$$

is a positive real number.

We chose an arbitrary member k of $T(Q)$, and we need to show that $k + 1$ is also a member. So, let $x_1, x_2, \ldots, x_k, x_{k+1}$ be any $k + 1$ positive real numbers. By the associative property,

$$\sum_{i=1}^{k+1} x_i = x_1 + x_2 + \ldots + x_k + x_{k+1}$$

$$= (x_1 + x_2 + \ldots + x_k) + x_{k+1}.$$

Since $k \in T(Q)$, we have that $\Sigma_{i=1}^{k} x_i$ is a positive real number and thus

$$(x_1 + x_2 + \ldots + x_k) + x_{k+1}$$

$$= \left(\sum_{i=1}^{k} x_i \right) + x_{k+1}$$

is the sum of *two* positive real numbers. Since $2 \in T(Q)$, we have that

$$\sum_{i=1}^{k+1} x_i \in T(Q).$$

I.e. the successor of every member of $T(Q)$ is also in $T(Q)$.

But then $T(Q)$ is a set which satisfies (i) and (ii) of the induction theorem, and therefore it contains all positive integers, i.e.

$$T(Q) = \mathbb{P}.$$

The proof is complete. ∎

The phrase "let $k \in T(Q)$," which we used at the start of the proof, is often called the **inductive hypothesis,** or the **hypothesis of induction.** The proof just presented was **by induction on the number of** addends, and the number k of addends was the **variable of induction.**

The principle of induction just illustrated as a method of proof is called the first principle of induction, to distinguish it from its equivalent variant in Theorem 16.5, below. In its *set version*, it appeared as the Induction Theorem 16.3 in the preceding section. In the following, we repeat this version and present its more ordinary form, the *proposition version,* as well.

16.4 The First Principle of Induction:
(*Set version*): Let $S \subseteq \mathbb{P}$ and let

  (i) *(basis)*: $1 \in S$, and
  (ii) *(induction step)*: $\forall k \in \mathbb{P}, (k \in S \Rightarrow k + 1 \in S)$.

   Then $S = \mathbb{P}$. ∎

(*Proposition version*): Let Q be a proposition of the form

$$Q: \ \forall n \subseteq \mathbb{P}, R(n) \text{ is true.}$$

Then, to prove Q, it is sufficient to prove both of the following:

  (i) *(basis)*: $R(1)$ is true (i.e., $R(n)$ is true when $n = 1$); and
  (ii) *(induction step)*: $R(k) \Rightarrow R(k + 1)$ (i.e., if $n = k$ is a positive integer for which $R(n)$ is true, then $R(n)$ is also true when $n = k + 1$. ∎

The second principle of induction is equivalent to the first principle of induction; that is, each may be proven from the other and therefore they may be used interchangeably. This second principle of induction is often called **weak induction,** since its inductive hypothesis (ii) is stronger than that of the first principle of induction, and hence the entire implication is stated in a weaker format. (The two principles actually have equal strength, since they are equivalent.) For that reason, the first principle of induction is called **strong induction.** However, these designations may cause confusion and, indeed, some authors call the second principle of induction "strong induction" because of the stronger premise. To avoid such confusion, we adhere to the present custom and identify the two principles as *first* and *second*. The reader may also find the second principle of induction identified as **complete induction.**[1]

16.5 The Second Principle of Induction:
(*Set version*): Let $S \subseteq \mathbb{P}$ and let

(i) (*basis*): $1 \in S$, and
(ii) (*induction step*):
$\forall k \in \mathbb{P}, (\{1, 2, \ldots, k\} \subseteq S \Rightarrow k + 1 \in S)$.

Then $S = \mathbb{P}$. ∎

(*Proposition version*): Let Q be a proposition of the form

$$Q: \ \forall n \in \mathbb{P}, R(n) \text{ is true.}$$

Then, to prove Q, it is sufficient to prove both of the following:

(i) (*basis*): $R(1)$ is true (i.e., $R(n)$ is true when $n = 1$); and
(ii) (*induction step*): Let k be such a positive integer that $R(n)$ is true *in each case* when $n \leq k$; then $R(k + 1)$ is also true. ∎

The difference between the two induction principles is that the premise in the induction step of the second has *all* positive integers less than $k + 1$ in S, while in the first only k is assumed in S; the consequence, $k + 1 \in S$, is provable on either premise. As was mentioned above, the two principles may be used interchangeably, but there are many instances in which one is preferable to the other.

Another example of a detailed proof by induction is given in the next section, where double induction is discussed. However, before leaving the elementary aspects of induction,

---

[1]Complete induction should not be confused with "perfect induction," which is a method of proving a proposition to be a tautology akin to proof by truth-tables.

we summarize its use (we chose the first principle) as a method of proof. The theorem to be proved is of the form "$\forall n \in \mathbb{P}$, $R(n)$ is true;" that is, an open sentence $R(n)$ concerning the (arbitrary) positive integer n, quantified universally.

To prove Q true, $R(n)$ must be proved true for every positive integer n. The proof by induction then calls for proving both of the following parts:

(i) (*Basis*): $R(1)$ is true; and

(ii) (*Induction step*): For each positive integer k, the assumption that $R(k)$ is true (called the **induction hypothesis**) implies that $R(k + 1)$ is true.

If these two parts are successfully proven, then the conclusion is that $R(n)$ is true for all positive integers n; i.e., that Q is true.

One last remark about proof by induction: It may be desired to prove that $R(n)$ is true for all positive integers n, beginning with, say, 6 instead of 1. Then, we simply replace the basis by:

(i) (*Basis*): $R(6)$ is true.

We also replace the induction step by:

(ii) (*Induction step*): The assumption that $R(k)$ is true, for some positive integer $k \geq 6$, implies that $R(k + 1)$ is true.

For the second principle of induction, the phrase "for some $k \geq 6$" in the induction step is replaced by, "for all $n \in \{6, \ldots, k\}$, for some positive integer $k \geq 6$."

16.12 DOUBLE INDUCTION

When the statement to be proved involves *two* arbitrary positive integers, the technique of proof is a variation on induction, called **double induction.**

In the following example, we assume that associativity of addition of positive integers has been proven. Examine the proposition R that addition of positive integers is commutative:

$$R:\ m + n = n + m,\ \forall m, n \in \mathbb{P}.$$

This statement may be proven in *two stages*. In the first stage, we prove that V is true, where V is:

$$V:\ m + 1 = 1 + m,\ \forall m \in \mathbb{P}.$$

Statement V is of the form

$$\forall m \in \mathbb{P},\ W(m)$$

and the usual induction argument suffices to prove it:

Basis: $W(1)$ is true, since $1 + 1$ does not distinguish between the two 1's.

Induction step: Suppose that $W(k)$ is true for some positive integer k (i.e., k is a member of the set $\{n \in \mathbb{P}: W(n)$ is true$\}$). Then, the *inductive hypothesis* is:

$$k + 1 = 1 + k.$$

Now, $W(k + 1)$ should be implied by $W(k)$, and so we had better *preview* the appearance of $W(k + 1)$:

$$(k + 1) + 1 = 1 + (k + 1).$$

To prove this, we start with the left-hand side of $W(k + 1)$ and convert it by *legitimate means* to the right-hand side of $W(k + 1)$:

$$(k + 1) + 1 = (1 + k) + 1, \qquad \text{by the inductive hypothesis}$$

$$= 1 + (k + 1), \qquad \text{by associativity of addition.}$$

We just proved $W(k + 1)$ and hence $W(k)$ does imply $W(k + 1)$.

We may now conclude (by induction) that $W(m)$ is true for all positive integers m, i.e. that V is true.

In the second stage, we build on V, which we have just proved:

$$V:\ m + 1 = 1 + m,\ \forall m \in \mathbb{P}.$$

We now wish to prove the original proposition R:

$$R:\ m + n = n + m,\ \forall m, n \in \mathbb{P},$$

but we now take R as a statement concerning the variable n:

$$Y: \forall n \in \mathbb{P}, [m + n = n + m, \forall m \in \mathbb{P}].$$

This statement is of the form:

$$\forall n \in \mathbb{P}, Z(n),$$

where $Z(n)$ is:

$$m + n = n + m, \forall m \in \mathbb{P}.$$

Again, we employ the usual induction argument:

Basis: $Z(1)$ is $m + 1 = 1 + m, \forall m \in \mathbb{P}$; it is precisely the statement V we just proved in the first stage of this double induction proof.

Induction step: Suppose, as the *inductive hypothesis*, that $Z(k)$ is true; that is, that $Z(n)$ is true when $n = k$:

$$m + k = k + m, \forall m \in \mathbb{P}.$$

Since $Z(k + 1)$ should be implied by $Z(k)$, we *preview* the appearance of $Z(k + 1)$.

$$Z(k + 1): m + (k + 1) = (k + 1) + m, \forall m \in \mathbb{P}.$$

To prove $Z(k + 1)$ true, we start with its left-hand side and convert it to its right-hand side *by legitimate means*, the inductive hypothesis and associativity:

$$
\begin{aligned}
m + (k + 1) &= (m + k) + 1, && \text{by associativity} \\
&= (k + m) + 1, && \text{by the inductive hypothesis} \\
&= k + (m + 1), && \text{by associativity} \\
&= k + (1 + m), && \text{by } V, \text{ proved above} \\
&= (k + 1) + m, && \text{by associativity.}
\end{aligned}
$$

This proves $Z(k + 1)$ true and hence $Z(k)$ does imply $Z(k + 1)$.

The conclusion is that $Z(n)$ is true for all positive integers n, and that is precisely what Y and R are. We thus proved R by double induction.

The **principle of double induction** has its strong and weak versions, as well. In fact, there are cases where it is convenient to use strong induction in one phase of the proof and weak induction in the other.

A fairly convenient form of the principle of double induction is the following theorem. It is stated in the more ordinary form of induction and is easy to prove from the principle of induction.

> ### 16.6 Principle of Double Induction:
> Let $P(m,n)$ be a proposition about the positive integers m and n. To prove that $P(m,n)$ is true for all positive integers m and n, it is sufficient to prove either (i) and (ii), or (i) and (iii), of the following:
>
> *(i)* (*Basis*): $P(1,1)$ is true.
>
> *(ii)* (*Induction step for first principle*): For any positive integers j and k, the assumption that $P(j,k)$ is true implies that $P(j + 1,k)$ and $P(j,k + 1)$ are true.
>
> *(iii)* (*Induction step for second principle*): For any positive integers j and k, the assumption that $P(m,n)$ is true for all positive integers $m \leq j$ and $n \leq k$ implies that $P(j + 1,k)$ and $P(j,k + 1)$ are true.

Again, the initial proposition need not have $m = 1$ and $n = 1$. In fact, the proposition P may not even make sense for such values. The basis may involve any two positive integers a and b (i.e., $P(a,b)$ is true), and then the proposition proved is $P(m,n)$ for all positive integers $m \geq a$ and $n \geq b$. In that case, the obvious changes need to be made in the statement of the induction step.

16.13 INDIRECT PROOFS

As was remarked earlier, an implication (the classical form of a theorem: premise(s) $\Rightarrow$ conclusion) may be proved *indirectly* by proving its contrapositive (negation of conclusion $\Rightarrow$ negation of premise). This technique starts just like a proof by contradiction, but instead of seeking a contradiction it proceeds to prove the negation of the premise, *without having assumed the latter.*

A comparison between the two is more easily seen from

their formal statements:

To prove $P \Rightarrow Q$:
By contradiction: $\sim(P \ \& \ \sim Q)$.
By indirection: $\sim Q \Rightarrow \sim P$.

We illustrate by turning the tables and proving the well-ordering principle by induction, this time assuming Peano's axioms, of course. To remind the reader, the well-ordering principle for positive integers is:

"Every nonempty set of positive integers has a least member."

Let $T \subseteq \mathbb{P}$ and let T have no least member (negating the conclusion). Then *define* the set S to include a positive integer n iff every member of T is greater than it; i.e.

$$S = \{n \in \mathbb{P}: x \subset T \Rightarrow x > n\}.$$

Now, $1 \in S$ because it cannot belong to T. (If $1 \in T$, then 1 is the least member of T since it is the least member of its superset $\mathbb{P}$. Yet T has no least member.)

Let $k \in S$. (We know $1 \in S$, so we may.) Then, by the definition of S, $\forall x \in T$, $x > k$ and thus $x \geq k + 1$. But if x *equals* $k + 1$ ($x = k + 1$), it must be the least member of T, since $k \in S$ and *every* $x \in T$ is greater than k. Thus $x \neq k + 1$ and it must be true that $x > k + 1$. But then every member x of T is greater than $k + 1$ as well, which makes $k + 1$ a member of S, by the definition of S. We showed that $k \in S \Rightarrow k + 1 \in S$.

By the induction theorem, $S = \mathbb{P}$, which puts all positive integers into S and leaves none for T, because now any member of T must be greater than the members of S, i.e. all the positive integers, and thus cannot be itself a positive integer. Hence $T = \phi$.

We proved for any $T \subseteq \mathbb{P}$:

$(T$ has no least member $\Rightarrow T = \phi)$

which is the contrapositive of, and thus equivalent to,

$(T \neq \phi \Rightarrow T$ has a least member$)$.

And *that* is the well-ordering principle. ∎

In proofs by induction, care should be taken to show that *both* conditions of the induction theorem are satisfied. There is a tendency to forget to show that 1 is in the set of positive integers for which the proposition is true. As a result, false statements are often "proved" in this fashion.

16.14 PROOF OF STATEMENT OF EQUIVALENCE

Often, one encounters a theorem whose statement starts with,

"The following conditions are (mutually) equivalent."

It then provides a list of conditions and the assertion is that *each is equivalent to each other.* If the list had just three conditions, call them (a), (b), and (c), what must be established one way or another is each of the bi-implications $(a) \iff (b)$, $(a) \iff (c)$, $(b) \iff (c)$. Since each bi-implication usually requires proving two implications, e.g. $((a) \iff (b)) \equiv ((a) \Rightarrow (b) \ \& \ (b) \Rightarrow (a))$, the task requires six separate proofs. Four equivalent conditions require twelve separate proofs, five require twenty, and so on.

The transitivity of equivalence makes it possible to prove only as many implications as there are equivalent conditions. (In practice, it may on occasion be easier to vary slightly from a strict application of this principle.)

A glance at the **circle of implications**

should suffice to convince the reader that all six implications are present. For example, $(c) \Rightarrow (b)$ is derived from $(c) \Rightarrow (a)$ and $(a) \Rightarrow (b)$ by transitivity of implication.

In the event that the prover of such a theorem finds it more convenient to create two (or more) implication circles, they must be connected once in each direction, as is illustrated by the two circles connected with $(b) \Rightarrow (d)$ and $(g) \Rightarrow (c)$:

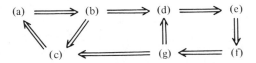

In fact, there are more implications here than are needed to assert equivalence of all parts. The reader may wish to start eliminating one implication at a time until the minimal number is reached.

17

Recursion

17.1 RECURSIVE DEFINITION OF FUNCTIONS

[*In this chapter, the domain of a function is either the set* $\mathbb{P}$ *of positive integers or the set* $\mathbb{N}$ *of nonnegative integers, unless otherwise indicated.*

With the domain understood and the co-domain implied by the "function rule," we resort to common usage and let the "definition of a function" mean the rule by which the function value is determined, rather than the more precise (and correct) collection of ordered pairs.]

Consider the function f defined by

$$f(n) = 2^n, \forall n \in \mathbb{N}.$$

We can list the values of f for the first few elements of $\mathbb{N}$:

$$f(0) = 2^0 = 1, \; f(1) = 2^1 = 2, \; f(2) = 2^2 = 4, \text{ etc.}$$

Just as well, we can compute the value, say, $f(6)$ as

$$f(6) = 2^6 = 64.$$

The function f was thus defined **explicitly** in terms of an operation (exponentiation) which can be employed directly.

Instead, let us define a function g as

$$\begin{cases} \text{(i)} \ \ g(0) = 1, \\ \text{(ii)} \ \ g(n + 1) = 2 \cdot g(n). \end{cases}$$

(It is to be understood that the second line carries the quantification "for all $n \in \mathbb{N}$ other than 0.")

At first glance, this definition may look improper: the function g is defined in terms of the function g itself. It is however *not* a circular definition. Part (ii) embodies the critical attitude which makes recursive definitions and recursive algorithms so useful. It says, *"If I knew what g(n) is, I could tell what g(n + 1) is."* It has the effect of *"passing the buck,"* of taking care *now* of one thing only—*how to compute the new value from old values.* It does, however, rely on the ability to obtain the old values when we need them, and that is very important. This is the purpose of part (i) of the definition of g: as we keep "passing the buck" further down the line, there has to be a stopping place, a sign which says, as it were, "the buck stops here." That happens when the value of g at 0 is given as 1: $g(0) = 1$ and the computation finally gets to that point.

To illustrate, when asked for $g(6)$ we can simply state that

$$g(6) = 2 \cdot g(5)$$

and leave it at that. However, this does not give a *number* as the immediate answer; it just delays the process of evaluation. To obtain a value, we must proceed as follows:

$$g(6) = 2 \cdot g(5) = 2 \cdot 2 \cdot g(4) = 2 \cdot 2 \cdot 2 \cdot g(3)$$
$$= 2 \cdot 2 \cdot 2 \cdot 2 \cdot g(2)$$
$$= 2 \cdot 2 \cdot 2 \cdot 2 \cdot 2 \cdot g(1)$$
$$= 2 \cdot 2 \cdot 2 \cdot 2 \cdot 2 \cdot 2 \cdot g(0).$$

Now, part (i) gives the value $g(0) = 1$ and hence we can obtain

$$g(6) = 2 \cdot 2 \cdot 2 \cdot 2 \cdot 2 \cdot 2 \cdot 1 = 2^6 = 64.$$

Of course, if we could afford to be very orderly, and possibly wasteful, about the process and generate the values of g at 0, 1, 2, etc. (without knowing how many such values we shall need), we would have $g(5)$ ready for use and could determine $g(6)$ immediately:

$$g(0) = 1;$$

$$g(1) = 2 \cdot g(0) = 2 \cdot 1 = 2^1 = 2;$$

$$g(2) = 2 \cdot g(1) = 2 \cdot 2 = 2^2 = 4;$$

$$g(3) = 2 \cdot g(2) = 2 \cdot 4 = 2^3 = 8;$$

$$g(4) = 2 \cdot g(3) = 2 \cdot 8 = 2^4 = 16;$$

$$g(5) = 2 \cdot g(4) = 2 \cdot 16 = 2^5 = 32.$$

And then, in one step,

$$g(6) = 2 \cdot g(5) = 2 \cdot 32 = 2^6 = 64.$$

The function g is identical to the function f, above. However, the value of g at an argument is specified in terms of values of *the same function g* for smaller arguments (except at 0, of course). The definition of g employs the *evaluating mechanism* of g itself to produce new values from old ones. This is what is called a **recursive definition** of a function. It consists of two parts:

(i) The **basis** or **initial condition,** which is a given value of the function at an argument; and

(ii) the **recursive clause** or **generating rule,** wherein the value of the function at an argument is defined in terms of the value(s) of the same function at smaller argument(s).

The initial condition of part (i) is often called the **termination condition** when the function is evaluated in decreasing order of the arguments.

Clear distinction should be made between proper and improper definitions of objects in terms of themselves. On the one hand, the recursive definition just presented always yields a unique value for g, with no ambiguity. The value of g we seek at $n + 1$ is defined in terms of *well-determined* values of g at smaller arguments—values we compute before we finish the computation of $g(n + 1)$.

On the other hand, let us "define" a (nonexistent) "compulsive set" as a set "in which each subset is compulsive." The "definition" leaves no way to determine what a "compulsive set" is, because it is a **circular definition,** one *whose meaning must already be known in order to be understood to begin with.* This, of course, is unacceptable.

This "definition" is somewhat similar to what happens when we give as the "definition" of a function h just the following:

$$h(n + 1) = 2 \cdot h(n).$$

To compute $h(6)$, we could try to proceed as we did with $g(6)$:

$$h(6) = 2 \cdot h(5) = \ldots = 2 \cdot 2 \cdot 2 \cdot 2 \cdot 2 \cdot 2 \cdot h(0).$$

But, since no value was given for $h(0)$, we must stop without an answer, or else we continue indefinitely after we leave the domain of nonnegative integers.

A **termination condition** is needed for the recursive definition to be a good definition. Suppose it is given as $h(0) = 1$; then we have $h = g = f$. For the sake of contrast, let us have instead a different basis, or termination condition, so that h is defined by

$$\begin{cases} \text{(i) } h(0) = 3, \\ \text{(ii) } h(n + 1) = 2 \cdot h(n). \end{cases}$$

Then, $h(1) = 2 \cdot 3 = 6, h(2) = 2 \cdot 6 = 12$, etc.

When the values of the function are computed in increasing order of the arguments, what was the *termination condition,* (i) $h(0) = 3$, in the opposite direction is the **initial condition** or the **boundary condition.** The second condition, (ii) $h(n + 1) = 2 \cdot h(n)$, the recursive focus of the definition, is variously called the **generating condition,** the **generating rule,** the **recursive step** or **clause** and the **inductive step.**[1]

A recursive definition of a familiar function often has an unfamiliar appearance. For instance,

---

[1] The term "inductive step" or **"induction step"** raises objections, since it tends to obscure the difference between recursion and induction, as discussed in Section 17.8, below. However, its users justify it by its similarity to the inductive process. When the generating rule is called the "induction step," the initial condition is almost always called the **basis.**

EXAMPLE 1:

The recursively defined function

$$\begin{cases} (i) \ j(0) = 3, \\ (ii) \ j(n + 1) = 2 \cdot j(n) - 3, \end{cases}$$

may not be easy to recognize as the constant function also definable by $j(n) = 3$.

EXAMPLE 2:

The recursively defined function

$$\begin{cases} (i) \ f(0) = 1, \\ (ii) \ f(n + 1) = (n + 1) \cdot f(n), \end{cases}$$

is probably easier to recognize as the **factorial** of n, also denoted by $f(n) = n!$

EXAMPLE 3:

The recursively defined function

$$\begin{cases} (i) \ a(0) = 5, \\ (ii) \ a(n + 1) = a(n) + 1, \end{cases}$$

may still be recognizable as addition to 5: $a(1) = a(0) + 1 = 5 + 1 = 6; a(4) = a(3) + 1 = a(2) + 1 + 1 = a(1) + 1 + 1 + 1 = a(0) + 4 = 5 + 4 = 9.$

EXAMPLE 4:

We assume that we know the names of members of $\mathbb{N}$ and the order in which they appear. We denote by y' the **successor** $y + 1$ of the nonnegative integer y.

We can now define addition (for a fixed value of x) recursively in a manner similar to Example 3, as follows:

$$\begin{cases} (i) \ x + 0 = x, \\ (ii) \ x + y' = (x + y)'. \end{cases}$$

These two equations *define* addition to x recursively. If $x = 5$, we have

$$5 + 0 = 5;$$

$$5 + 1 = 5 + 0' = (5 + 0)' = 5' = 6;$$

$$5 + 2 = 5 + 1' = (5 + 1)' = 6' = 7;$$

etc.

EXAMPLE 5:

In a manner similar to Example 4, multiplication (for a fixed value of x) may be defined recursively by,

$$\begin{cases} (i) \ x \cdot 0 = 0, \\ (ii) \ x \cdot y' = x \cdot y + x. \end{cases}$$

We can now find successive values of $x \cdot y$, with $x = 3$ as an example:

$$3 \cdot 0 = 0;$$
$$3 \cdot 1 = 3 \cdot 0' = 3 \cdot 0 + 3 = 0 + 3 = 3;$$
$$3 \cdot 2 = 3 \cdot 1' = 3 \cdot 1 + 3 = 3 + 3 = 6;$$

etc.

The reader may wish to consider at this point the familiar definitions of general addition and multiplication. Their recursive definitions, with the use of primitive recursion, are given in Section 17.9, below.

EXAMPLE 6:

Some functions are easy to define recursively and more difficult to define otherwise. For instance, the function $F: \mathbb{N} \to \mathbb{N}$ whose successive values are the **Fibonacci numbers** is defined recursively by,

$$\begin{cases} F(0) = 0; \\ F(1) = 1; \\ F(n + 1) = F(n) + F(n - 1), \forall n \in \mathbb{P}. \end{cases}$$

To find a new Fibonacci number, simply add the last two:

$$F(2) = F(1) + F(0) = 1 + 0 = 1;$$
$$F(3) = F(2) + F(1) = 1 + 1 = 2;$$
$$F(4) = F(3) + F(2) = 2 + 1 = 3;$$
$$F(5) = F(4) + F(3) = 3 + 2 = 5;$$

etc. The list of the first few Fibonacci numbers may be easily computed as $0, 1, 1, 2, 3, 5, 8, 13, \ldots$

17.2 RECURSIVE SOLUTIONS TO PROBLEMS; TOWERS OF HANOI

An important feature of a recursive definition of a function can be made clear when a solution is needed to a complicated problem. We take as an example the well known Towers-of-Hanoi Puzzle, illustrated in Figure 17.1.

The task is to move all the discs from peg A to peg C, one at a time. In each move, the top disc is removed from a peg and placed on another peg, either on the bottom or on a larger disc. We must find the smallest number of moves required to thus move n discs. In the process, we should also find a method for accomplishing the task.

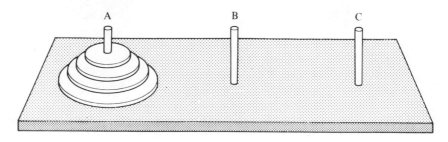

The Towers-of-Hanoi Puzzle, View I

FIGURE 17.1

The recursive approach to solving this problem restricts the attention to just one step: If we knew how to move $n - 1$ discs in the required manner from one disc to another, *what else* do we need to do to move all n discs?

Suppose we know how to move the top three discs (in Figures 17.1 and 17.2) from peg A to peg B, temporarily. Then, after doing so (in a way we have not yet revealed), we move the bottom disc from peg A to peg C, and then we repeat the process of moving the three discs, only that now we move them from peg B to peg C. Then the puzzle is complete.

We thus dealt with the current step only, leaving the solution of previous steps for later.

Now, how many moves are required to complete the entire operation? Let us denote the number of moves needed to transfer n discs from one peg to another by $H(n)$. Then, we employed the following number of moves:

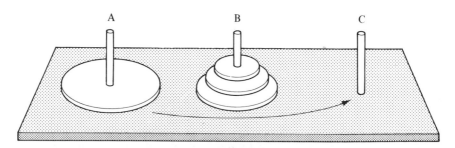

The Towers-of-Hanoi Puzzle, View II

FIGURE 17.2

$H(3)$, to move the top three discs from A to B,

1, to move the bottom disc from A to C,

$H(3)$, to move the three discs from B to C.

Then, $H(4) = 2 \cdot H(3) + 1$.

The same approach will work when we wish to transfer any number n of discs from one peg to another. The **recursive relation,** then, is

$$H(n + 1) = 2 \cdot H(n) + 1.$$

All that we need now is the *termination condition,* which clearly is

$$H(1) - 1.$$

That is, one disc can be transferred to another peg in one move. We may now compute

$$H(4) = 2 \cdot H(3) + 1$$
$$= 2 \cdot (2 \cdot H(2) + 1) + 1$$
$$= 2 \cdot (2 \cdot (2 \cdot H(1) + 1) + 1) + 1$$
$$= 2 \cdot (2 \cdot (2 \cdot 1 + 1) + 1) + 1$$
$$= 15.$$

Not only can we compute the value of the function H at each n, but we can also reconstruct the individual moves: first moving the top disc, then the top two discs, next the top three, and so forth. (The reader may wish to try out this sequence.)

The truly remarkable feature of the recursive solution is that it concerns only **reducing the problem** from $n + 1$ discs to the same problem with n discs. The rest took care of itself.

17.3 RECURSIVE DEFINITION OF STRUCTURES

The same principle we used in the preceding sections may be applied to structures other than functions, provided the structures possess stages whose dependence on previous stages is sufficiently uniform in nature. We illustrate with the definition of a **rooted tree** (Definition 6.28).

We could regard the root of the tree as the 0-th stage, or as a rooted tree of **height** 0. The set of vertices of the first stage, called V_1, consists of all the terminal vertices of edges whose initial vertex is the root of the entire tree. Each vertex in stage 2 is the terminal vertex of exactly one edge, and the initial vertex of that edge is a vertex in stage 1. The set of vertices of stage 2 is called V_2. And so on.

> *17.1 Recursive Definition:*   A directed graph $G = (V,E)$ is a **rooted tree of height** h ($h \geq 0$) iff the set of vertices V can be partitioned recursively into (disjoint) sets $V_0, V_1, \ldots, V_h$ so that:
>
> *(i)* V_0 has a single vertex, called the **root,** which is the terminal vertex of no edge.
> *(ii)* For each $i \in \{0,1, \ldots, h - 1\}$, V_{i+1} is a finite nonempty subset of V such that every vertex v in V_{i+1} is the terminal vertex of exactly one edge, and the initial vertex of that edge is a vertex in V_i.

The above example is typical of recursive definitions of structures. At times, the terseness of such definitions may damage clarity or speed of understanding to some degree, but the statements of such definitions are conveniently short and are usually easy to follow in construction or verification and identification.

17.4 RECURSIVE ALGORITHMS; BINARY-TREE TRAVERSAL

In Section 6.16, three orders of traversal were given for a binary tree: pre-order, in-order and end-order. The traversal algorithms were described in detail and at length. However, each traversal order may be described recursively in a briefer statement, as is seen below. In the following definitions, each vertex is regarded as the root of the binary tree which originates with it.

17.2 Recursive Definition of Pre-Order Traversal of a Binary Tree:

(i)   Visit the root.
(ii) Visit the left subtree in pre-order traversal.
(iii) Visit the right subtree in pre-order traversal.

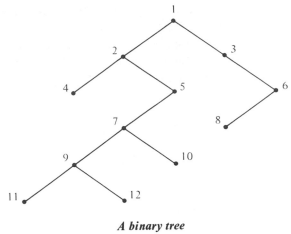

A binary tree

FIGURE 17.3

We follow the instructions given in the recursive definition on the binary tree of Figure 17.3. The root 1 is visited first, and then the left subtree (with root 2) must be visited in pre-order traversal. Thus, the right subtree (with root 3) will wait until the entire left subtree is traversed. Here, we apply the recursive process to the left subtree in its turn: we visit its root 2 and then its left subtree 4, before traversing the right subtree (with root 5) in pre-order.

If we follow the instructions given in the recursive definition, we would visit the vertices in the following order:

$$1,2,4,5,7,9,11,12,10,3,6,8.$$

17.3 Recursive Definition of In-Order[1] Traversal of a Binary Tree:

(i) Visit the left subtree in in-order traversal.

(ii) Visit the root.

(iii) Visit the right subtree in in-order traversal.

The order of visit of the vertices of the same binary tree of Figure 17.3 is:

$$4,2,11,9,12,7,10,5,1,3,8,6.$$

---

[1]In-order is sometimes called **post-order** and **symmetric-order.**

17.4 Recursive Definition of End-Order[2] Traversal of a Binary Tree:

(i) Visit the left subtree in end-order traversal.

(ii) Visit the right subtree in end-order traversal.

(iii) Visit the root.

The order of visit of the vertices of the binary tree of Figure 17.3 is:

$$4,11,12,9,10,7,5,2,8,6,3,1.$$

17.5 RECURSION IN PROGRAMMING

In many cases, the economy achieved by a recursive definition or a recursive specification is more in the statement than in the computation itself. This economy of statement, however, is of considerable value in programming, where human effort is increasingly more valuable when compared with machine effort.

Yet, the use of recursion in programs is often at the cost of serious expenditure of time and space, even when the programming language permits recursion. (In many programming languages, such as FORTRAN, subroutines cannot call themselves. When recursion is permitted, as in ALGOL, PASCAL, LISP, APL, PL/I, it is a powerful addition to a programming language.)

A **recursive routine** is a routine which uses itself as a subroutine; i.e., *a routine which calls itself.* A recursive routine must have a termination criterion to avoid an infinite loop of calling itself.

For example, a recursive routine to compute the **factorial** of *n* may appear as in Figure 17.4. This recursive routine is called first with the present value of *n*, say $n = 5$. Since $5 < 3$ is false, 5*FACTORIAL(4) is returned, but this cannot be evaluated until the value of FACTORIAL(4) is determined.

At this point, another call on the routine FACTORIAL is executed with $n = 4$ (and the original calculation of 5! is waiting unfinished). This process continues with another (interrupting) call to the routine to find FACTORIAL(3), and another yet to find FACTORIAL(2). Since FACTORIAL(2)

---

[2]End-order is sometimes called **post-order** and **bottom-up order**.

Procedure FACTORIAL (n)
   if n < 3 then return n;
     else return (n * FACTORIAL (n−1));
   end if;
end procedure;

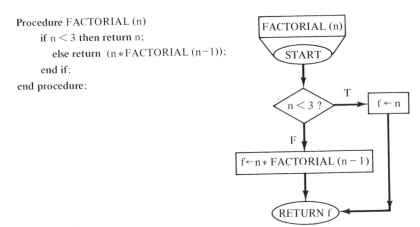

Recursive procedure for factorial n

FIGURE 17.4

is given as 2, the calculation of FACTORIAL(3) can now resume and, when it is completed, the waiting calculation of FACTORIAL(4) can be completed. Only then can the original interrupted computation of FACTORIAL(5) be concluded.

The implementation of such a programming device must make room *during execution* for the interrupted computations, and it must link them properly, so that the eventual calculation can be completed correctly.

17.6 RECURSION AND ITERATION

The same result obtained by the recursive calculation of the factorial function of the last section could be achieved by iteration (usually at less cost of both time and space during execution, but at greater cost to the programmer), as in Figure 17.5.

A similar relationship between recursion and iteration may be seen in the computation of the Fibonacci numbers. For convenience, we recast the Fibonacci numbers (Example 6, Section 17.1) so that the *first* is $F(1) = 0$, the second is $F(2) = 1$, etc. The recursive treatment may then appear as in Figure 17.6.

By comparison, the iterative routine may appear as in Figure 17.7.

Procedure ITFACT (n)
    value ← 1; i ←1;
    while i < n do
        value ← i * value;
        i ← i + 1;
    end while;
    return value;
end procedure;

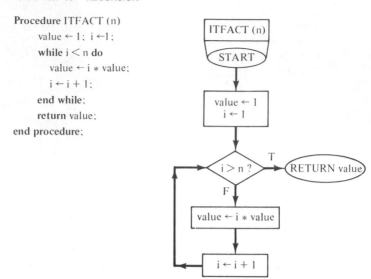

Iterative procedure for factorial n

FIGURE 17.5

Procedure FIBONACCI (n)
    if n < 3 then return n − 1;
        else return (FIBONACCI (n − 1) +
                FIBONACCI (n − 2));
    end if;
end procedure;

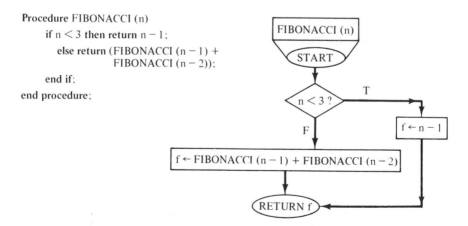

Recursive procedures for Fibonacci numbers

FIGURE 17.6

Perhaps the essential difference between the iterative approach and the recursive one may be seen in the direction of computation:

To iterate, start with the initial condition and build up the values for increasingly larger arguments; at each stage, all needed values of variables *are available for immediate use.*

In the recursive approach, instructions are given for obtaining new values from old. The old values may not be ready when needed and thus may have to be computed before the new value can be determined. During the computation of the old values, they themselves are regarded as new values which depend on old values not yet computed; and so on until the termination condition is reached.

Procedure ITFIB (n)
    if n < 3 then return n − 1;
      else i ← 3
        first ← 0;
        second ← 1;
        while i ≤ n do
          temp ← second;
          second ← second + first
          first ← temp;
          i ← i + 1;
        end while;
        return second;
    end if;
end procedure;

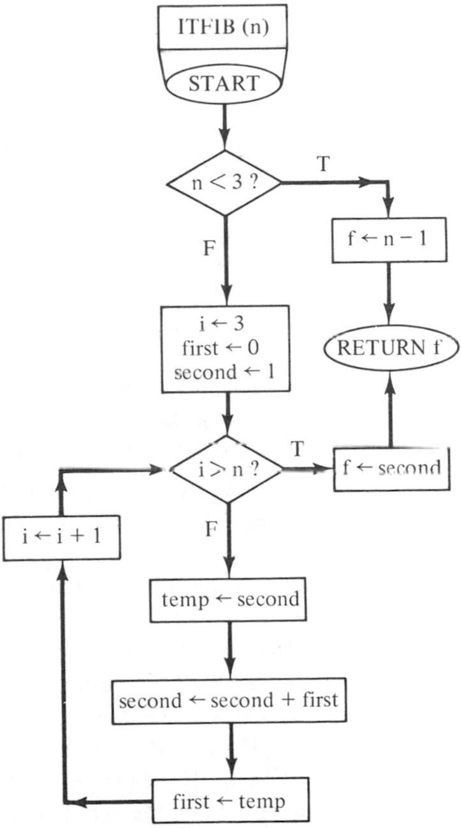

Iterative procedure for Fibonacci numbers

FIGURE 17.7

17.7 RECURSIVE DEFINITION OF SETS AND FORMAL SYSTEMS

Definition 4.26 of the **powers of a set** is recursive, since it defines the $n + 1$st power of a set in terms of its n-th power. (In it, we rely on the product of sets from Definition 4.24.) We repeat the definition here for the reader's convenience.

> ***17.5 Definition:*** Let A be a set. Then,
>
> *(i)* $A^0 = \Lambda = \{\lambda\}$.
> *(ii)* $A^1 = A$.
> *(iii)* $A^{n+1} = A^n \cdot A$.

Such a definition is very similar to the recursive definition of a function: it has a basis (parts (i) and (ii)) and a recursive-generating step (iii). However, in many instances, especially when a set is specified in terms of membership in it, its recursive definition has an additional requirement, as may be seen from the following example.

EXAMPLE 1:

Define a set S of (real) numbers as follows (addition of real numbers is assumed):

> ***Rule 1:*** $1 \in S$.
>
> ***Rule 2:*** If x and y are (not necessarily distinct) members of S, then $x + y \in S$.
>
> ***Rule 3:*** No number is a member of S unless it is the result of a finite number of uses of rules 1 and 2.

It should not be difficult to identify the set S as the set $\mathbb{P}$ of all positive integers. In the definition, each of the three rules plays a crucial role, with the first two being the familiar ones. Rule 1 is called the **basis.** It identifies the **primitive,** or original, members of the set, the members which are used to generate other members of the set.

Rule 2 is called the **generating rule,** for the obvious reason—it tells how to get new members from old.

Rule 3 is called the **disclaimer** or the **extremal clause.** Without it, there may be unwanted numbers included as members of the set. (This is a point not to be overlooked: the generating rule guarantees membership to the desired numbers, but *it does not prohibit* membership of undesired members; that has to be done separately by the disclaimer in the third rule.)

Section 4.12 on closure of families of sets, and Section 4.13 on defining sets by closure, furnish examples of recursive definitions of sets. Following are several examples of such definitions.

EXAMPLE 2:

Let $\Sigma = \{0,1\}$. Define the **(Kleene) closure** Σ^* of Σ by:

(i) (Basis): $\lambda, 0, 1 \in \Sigma^*$.

(ii) (Generating Rule): If $x \in \Sigma^*$, then the two concatenations $x0$ and $x1$ are also members of Σ^*.

(iii) (Disclaimer): Nothing is a member of Σ^* unless it is the result of a finite number of uses of rules (i) and (ii).

This is an example of defining a set by closure—the concatenation-closure over Σ. It is, of course, of a recursive nature, since it uses elements already known to be in the set in order to identify new elements in the set.

EXAMPLE 3:

In Section 4.13 there are given two subsets:

$A = \{1,2,3\}$ and $B = \{2,4\}$ of the universal set $U = \{1,2,3,4\}$.

The family $G = \{A,B\}$ is the *basis,* and several of its closures are shown. These are closures under (1) finite unions, (2) finite intersection, (3) both finite unions and finite intersections, (4) relative complements, (5) both finite unions and relative complements, (6) complements, and (7) both finite unions and complements. Each of these definitions has a recursive nature:

Let the first rule always be,

(i) $A,B \in G^*$.

The disclaimer always prohibits any set from membership, with the exception of the ones specified by the first two rules. Then, the following generating rules produce the desired closures, respectively:

$(ii)_1: \forall x,y \in G^*, x \cup y \in G^*$.

$(ii)_2: \forall x,y \in G^*, x \cap y \in G^*$.

$(ii)_3: \forall x,y \in G^*, x \cup y \in G^*$ and $x \cap y \in G^*$.

$(ii)_4: \forall x,y \in G^*, x - y \in G^*$.

$(ii)_5: \forall x,y \in G^*, x \cup y \in G^*$ and $x - y \in G^*$.

$(ii)_6: \forall x \in G^*, x' \in G^*$.

$(ii)_7: \forall x,y \in G^*, x \cup y \in G^*$ and $x' \in G^*$.

In each case, members of G^*, other than those in the basis, were specified in terms of members already known to be in G^*; that is the nature of the recursive definition of a set.

As was mentioned above, the disclaimer for each case is,

(iii) No other sets are in G^*.

The case would be far more interesting and instructive if U, A and B were infinite sets. Yet, the substance of the issue can be adequately captured from the simple cases of this example.

Formal systems often make use of permissible formulas, which are formed according to specified rules, called formation rules. The rules are frequently given recursively and may be regarded as rules for membership in the set of permissible formulas, rather than rules for forming the formulas.

For example, the correct formation of parenthesized expressions (only from the point of view of correct parenthesizing) may be given recursively. Usually, parentheses are used as grouping symbols, indicating precisely what part of the formula is taken by the operand or the argument.

EXAMPLE 4:

We ignore all symbols of the system which are neither ")"nor"(" and observe only the "correct" parenthesizing. The set *WFP* of well-formed-parenthesizing schemes is recursively defined by:

(i) *(Basis)*: ( ) $\in$ *WFP*.
(ii) *(Generating Rule)*: If $x,y \in$ *WFP*, then both (x) and xy are members of *WFP*.
(iii) *(Disclaimer)*: Nothing else is a member of *WFP*.

Starting from the basis of Example 4, and employing the generating rule in all possible ways, the first few resulting parenthesizing schemes are,

$$( )$$
$$(( )), ( )( )$$
$$((( ))), (( )( )), (( ))( ), ( )(( )), ( )( )( ).$$

It is not possible to obtain ( ))(( ), for example, which is improper parenthesizing.

EXAMPLE 5:

wff

The formation of the **well-formed-formulas (wff)** of the **propositional calculus** (of logic) may be specified recursively as follows.

EXAMPLE 5—*Continued*

Let $a, b, \ldots, z, a_1, b_1, \ldots, z_1, a_2, \ldots$ be an infinite list of propositional variables. (In the principal interpretation, these are variables whose range is $\{t, f\}$, for "true" and "false.")

(i) (*Basis*): Any propositional variable is a *wff*; also the primitive constant f is a *wff*.

(ii) (*Generating Rule*): If A and B are *wff*'s, then $[A \supset B]$ is a *wff*.

(iii) (*Disclaimer*): Only the results of a finite number of uses of rules (i) and (ii) are *wff*'s.

Examples of *wff*'s in this formulation of the propositional calculus are:

1. a.
2. f.
3. $[[[p \supset f] \supset f] \supset p]$.
4. $[[p \supset [q \supset s]] \supset [[s \supset p] \supset [s \supset q]]]$.

A student of logic might recognize the third example as an axiom of the propositional calculus (in its classical rendition); the first and the fourth are not theorems, and the second is a fallacy, or a contradiction. It is clear, then, that the rules in the example control only the *formation* of the well-formed formula; they tell whether a string of allowable characters is a formula which is even to be tested in the system. Whether such a *wff* is a theorem in the system, whether its value in the principle interpretation is always t, is not the concern of the formation rules in Example 5.

EXAMPLE 6:

A parallel development of the propositional calculus makes use of the **negation** symbol $\sim$ (or $\neg$) instead of the primitive constant f. (The list of propositional variables is the same as in Example 5.) The formation of well-formed-formulas is given recursively by the following:

(i) (*Basis*): A propositional variable is a *wff*.

(ii) (*Generating Rule*):
  1. If A is a *wff*, then $\sim A$ is a *wff*;
  2. If A and B are *wff*'s, then $[A \supset B]$ is a *wff*.

(iii) (*Disclaimer*): Nothing else is a *wff*.

There are, indeed, other formulations of the propositional calculus, and each has its own formation rules for the well-formed formulas in it.

EXAMPLE 7:

Regular expressions over an alphabet Σ are strings of symbols from among members of Σ and also from among $\varnothing$, λ, $\cup$, $\cdot$, *, (,). Each regular expression represents a set of strings of members of Σ which is recognizable by a finite-state acceptor. (For example, see Section 6.19, Figure 6.37.)

As was mentioned above, instead of stating formation rules for admissible formulas, the set of all such formulas may be defined recursively. We shall do so here with the **set $RE(\Sigma)$ of all regular expressions over Σ.**

(i) (*Basis*): 1. Every member of Σ is a member of $RE(\Sigma)$.
2–3. ϕ and λ are members of $RE(\Sigma)$.

(ii) (*Generating Rule*): 4–5. If x and y are in $RE(\Sigma)$ then $(x \cdot y)$ and $(x \cup y)$ are also in $RE(\Sigma)$.
6. If x is in $RE(\Sigma)$, then (x)* is also in $RE(\Sigma)$.

(iii) (*Disclaimer*): 7. Nothing is in $RE(\Sigma)$ unless it is the result of a finite number of applications of rules 1–6.

Regular expressions are an important part of finite automata theory, formal language theory, and several other areas in computer science. This exposition here clearly says very little about regular expressions and their use, just as Examples 5 and 6 say very little about the propositional calculus. These examples are presented here as illustrations of recursive definitions of sets and recursive specification of formation rules of formulas in formal systems.

17.8 RECURSION AND INDUCTION

The similarity between the generating rule in a recursive definition on the one hand and mathematical induction on the other is obvious. In fact, it is not uncommon to refer to a **recursive definition** as an **inductive definition.** Yet, there is a formal difference between the two which should be appreciated. Perhaps the following example may highlight the difference.

The principle of induction may be used to prove that a recursively defined function is **well-defined.** The question whether a function is well-defined arises when the definition of the function is perceived as inconclusive: If there is more than one way to compute $f(x)$ for a value of x, could two such different ways yield two different values for $f(x)$? If so, the so-called "definition" of the function is not a definition at all. (At least, it is not a definition of a function.)

For example, "define" $f: \mathbb{N}^2 \rightarrow \mathbb{N}$ by:

$$\begin{cases} \textbf{(a)} \ f(0,0) = 0; \\ \textbf{(b)} \ f(x,y + 1) = f(0,y) + x; \\ \textbf{(c)} \ f(x + 1,y) = f(x,0) + y. \end{cases}$$

If we compute $f(2,3)$ by rule (b), we get

$$f(2,3) = f(2,2 + 1) = f(0,2) + 2 = f(0,1) + 0 + 2$$
$$= f(0,0) + 0 + 0 + 2 = 0 + 2 = 2.$$

On the other hand, by rule (c) we have,

$$f(2,3) = f(1 + 1,3) = f(1,0) + 3$$
$$= f(0,0) + 0 + 3 = 0 + 3 = 3.$$

We then say that f, as given by the rules (a), (b) and (c), is **not a well-defined function.**

In contrast, consider $g: \mathbb{N}^2 \rightarrow \mathbb{N}$ "defined" by,

$$\begin{cases} \textbf{(d)} \ g(0,0) = 0; \\ \textbf{(e)} \ g(x,y + 1) = g(x,y) + 1; \\ \textbf{(f)} \ g(x + 1,y) = g(x,y) + 2. \end{cases}$$

It may appear that a conflict is possible for some order pair (x,y) among the rules (d), (e) and (f), just as was the case with the rules (a), (b) and (c). A few attempts at evaluating g may indicate otherwise:

$$g(0,1) = g(0,0) + 1 = 1.$$
$$g(1,0) = g(0,0) + 2 = 2.$$
$$g(1,1) = g(1,0) + 1 = 2 + 1 = 3; \text{ but also}$$
$$g(1,1) = g(0,1) + 2 = 1 + 2 = 3.$$

And so on.

It may well be that the rules (d), (e) and (f) do not cause a conflict, even though they have overlapping jurisdictions. If such is the case, g is a well-defined function, but such a claim requires proof.

It is often the case that, what appears to be a *recursive*

definition of a function, especially over a domain which is itself a recursively defined set, is suspect. Then, *induction* is used to prove the function well-defined.

*17.9 PRIMITIVE RECURSIVE FUNCTIONS

[*The main treatment of recursion for the computer scientist in this book ended in the preceding section. There are, however, concepts associated with recursion which may confront the reader and which are of deeper nature. They occur frequently in branches of logic and formal language theory. The account given them in this section, and the ones following, is terse by necessity—hardly more than a collection of definitions.*]

17.5 Definition: The **basic recursive functions** (on the set $\mathbb{N}$ of nonnegative integers) are:

1. The **successor function** s: $\mathbb{N} \rightarrow \mathbb{N}$ defined by,

$$s(x) = x + 1, \forall x \in \mathbb{N}.$$

(See the discussion of Peano's axioms in Section 15.2.)

2. The **zero function** z: $\mathbb{N} \rightarrow \mathbb{N}$ defined by,

$$z(x) = 0, \forall x \in \mathbb{N}.$$

3. The **(generalized) identity functions** (sometimes ca d the **selection functions**) for each i and n in $\mathbb{N}$ with $0 < i \leq n$, denoted by U_i^n : $\mathbb{N}^n \rightarrow \mathbb{N}$ and defined by,

$$U_i^n(x_1, x_2, \ldots, x_n) = x_i. \blacksquare$$

The selection function U_i^n selects as its image the i-th argument out of its n arguments. For example, the function U_2^3 is defined on the ordered triples of nonnegative integers and its image is always the second component:

$$U_1^3(3,4,5) = 3, \quad U_2^3(3,4,5) = 4, \quad U_3^3(3,4,5) = 5.$$

Also $U_1^2(6,9) = 6$ and $U_3^4(4,5,6,7) = 6.$

We now define two *operations* on functions which produce new functions. The first is the operation of **composition.** (Com-

---

*This section requires somewhat greater mathematical maturity.

position of functions was the subject of Section 9.4. Here, it is recast in a formal manner, to accommodate the development of the primitive recursive functions.) Consider, for example, the function $f: \mathbb{N}^2 \rightarrow \mathbb{N}$, defined by

$$f(x,y) = x + 3, \forall x,y \in \mathbb{N}.$$

Thus, $f(2,7) = f(2,172) = 5; f(5,9) = 5 + 3 = 8$; etc.

The function f may be obtained by composition of some of the basic recursive functions as follows: We begin with the selection function $U_1^2 : \mathbb{N}^2 \rightarrow \mathbb{N}$, defined by $U_1^2(x,y) = x, \forall x,y \in \mathbb{N}$. We can now obtain $s(U_1^2(x,y)) = x + 1, s(s(U_1^2(x,y))) = (x + 1) + 1 = x + 2$, and finally $f = s(s(s(U_1^2(x,y)))) = (x + 2) + 1 = x + 3$. Then, f is *a function whose argument has been replaced by a function* It then has been redefined by composition.

When g is a function of, say, *four* variables (i.e., $g: \mathbb{N}^4 \rightarrow \mathbb{N}$), a general instance of it may appear as $g(y_1,y_2,y_3,y_4)$. Let q, r, s and t be each a function of *three* variables (i.e., $q: \mathbb{N}^3 \rightarrow \mathbb{N}$, etc.). Now, suppose that when we evaluate the four functions at the ordered triple (1,7,9), we get $q(1,7,9) = 5$, $r(1,7,9) = 2, s(1,7,9) = 6$ and $t(1,7,9) = 3$. There is nothing then to prevent the evaluation of g at these four values, i.e. $g(5,2,6,3)$. In fact, we may write instead $g(q(1,7,9), r(1,7,9), s(1,7,9), t(1,7,9))$. Along these same lines, we could define a function h to be always evaluated in this fashion:

$$h(x_1,x_2,x_3) = g(q(x_1,x_2,x_3), r(x_1,x_2,x_3), s(x_1,x_2,x_3), t(x_1,x_2,x_3)).$$

We then say that h is defined by the operation of **composition** on the functions g, q, r, s and t.

Another example of a function defined by composition on the basic recursive functions is the following:

The constant function defined by $F(x) = 1$ may be obtained as the composition of the *successor function* and the *zero function*:

$$F(x) = s(z(x)) = s(0) = 1.$$

Similarly, the constant function defined by $T(x) = 2$ may be obtained by composition as

$$T(x) = s(F(x)) = s(s(z(x))).$$

17.6 *Definition:* Given the functions whose names, values and domains are shown by,

$$f(y_1, \ldots, y_k), \quad g_1(x_1, \ldots, x_n),$$

$$g_2(x_1, \ldots, x_n),$$

$$\ldots$$

$$g_k(x_1, \ldots, x_n).$$

Then the function c, defined below, is obtained from the above functions by composition:

$$c(x_1, \ldots, x_n) = f(g_1(x_1, \ldots, x_n), \ldots, g_k(x_1, \ldots, x_n)). \quad \blacksquare$$

The second operation on functions to be introduced here is called **primitive recursion.** It produces a new function from two given functions, and here, too, the sizes of the respective domains must fit precisely.

The basic idea is very similar to the recursive definition of functions. For example, we shall define a function we shall call the **predecessor function,** which yields as value an integer smaller by 1 than the argument (except at 0): $p \colon \mathbb{N} \to \mathbb{N}$ is defined as

$$p(x) = \begin{cases} x - 1, \text{ for } x \geq 1, \\ \quad 0 \;\; , \text{ for } x = 0. \end{cases}$$

To define p by primitive recursion, we *tell* its value at $x = 0$,

$$p(0) = 0.$$

Then we tell how to find $p(x + 1)$ if we know $p(x)$:

$$p(x + 1) = x, \; \forall x > 0.$$

The two important points to note are, first, that p is defined for $x + 1$, not for x as is the usual custom and, second, that we could have written

$$p(x + 1) = U_1^1(x)$$

expressing p in terms of the basic recursive functions.

As another example of the operation of primitive recursion, consider the function called **proper subtraction,** which is similar to the familiar subtraction, but which substitutes 0 for all negative results:

$$x \,\dot{-}\, y = \begin{cases} x - y, \text{if } x \geq y, \\ 0, \quad \text{otherwise.} \end{cases}$$

(For example, $6 \,\dot{-}\, 2 = 4$ but $2 \,\dot{-}\, 6 = 0$.) We recast proper subtraction so that it appears as a function, rather than an operation. D: $\mathbb{N}^2 \rightarrow \mathbb{N}$ is defined by,

$$D(x,y) = \begin{cases} x - y, \text{ if } x \geq y, \\ 0, \quad \text{otherwise.} \end{cases}$$

To define D by primitive recursion, we first specify its value when $y = 0$,

$$D(x,0) = x;$$

Then we instruct how to find $D(x,y + 1)$ if we know $D(x,y)$:

$$D(x,y + 1) = p(D(x,y)),$$

where p is the predecessor function just illustrated above. The two parts together constitute the definition of D by primitive recursion. (It is worth noting that, since the predecessor function p was defined by primitive recursion on basic recursive functions, D too is traceable to the basic recursive functions in the same manner.)

A few calculations illustrate the structure of the definition by primitive recursion:

$$2 \,\dot{-}\, 1 = D(2,1) = p(D(2,0)) = p(2) = 1.$$
$$2 \,\dot{-}\, 2 = D(2,2) = p(D(2,1)) = p(1) = 0.$$
$$2 \,\dot{-}\, 3 = D(2,3) = p(D(2,2)) = p(0) = 0.$$

The principle of primitive recursion is somewhat more general, since it applies to a function of any positive number of variables. The general pattern of primitive recursion is shown in the following definition by the two defining equations for the function R in terms of the two functions F and G.

17.7 Definition: Let $n > 1$ and let the functions $F: \mathbb{N}^{n-1} \to \mathbb{N}$ and $G: \mathbb{N}^{n+1} \to \mathbb{N}$ be given. Then

$$\begin{cases} R(0, x_2, \ldots, x_n) = F(x_2, \ldots, x_n), \\ R(x_1 + 1, x_2, \ldots, x_n) = G(x_1, R(x_1, x_2, \ldots, x_n), x_2, \ldots, x_n), \end{cases}$$

defines the function $R: \mathbb{N}^n \to \mathbb{N}$ by **primitive recursion** in terms of F and G. ∎

In this definition, the argument to which the primitive recursion is applied is the first; it could have been any of the n arguments, with the obvious corresponding changes.

It may help to note that, to obtain the value of $R(x_1 + 1, x_2, \ldots, x_n)$, we must know both $R(x_1, x_2, \ldots, x_n)$ and $G(x_1, y, x_2, \ldots, x_n)$, where $y = R(x_1, \ldots, x_n)$. As in the cases illustrated in earlier sections, such an attempt at evaluation will eventually reduce to the inquiry of the value $R(0, x_2, \ldots, x_n)$, which is defined in terms of $F(x_2, \ldots, x_n)$. The latter is a function of only $n - 1$ arguments, and the argument which is missing here, the one to be ignored, is precisely the subject of the definition by primitive recursion—the first one.

We present a few specific examples to illustrate:

1. **Addition:** $A(x, y) = x + y$.
   With primitive recursion:

$$\begin{cases} A(0, y) = y, \\ A(x + 1, y) = s(A(x, y)) = A(x, y) + 1. \end{cases}$$

Here $R(x_1, \ldots, x_n) = A(x_1, x_2)$,

$$A(0, x_2) = F(x_2) = U_1^1(x_2) = x_2,$$

$$A(x_1 + 1, x_2) = G(x_1, R(x_1, x_2), x_2) = G(x_1, A(x_1, x_2), x_2)$$

$$= s(U_2^3(x_1, A(x_1, x_2), x_2)) = s(A(x_1, x_2)) = A(x_1, x_2) + 1.$$

2. **Multiplication:** $M(x, y) = x \cdot y$.
   With primitive recursion:

$$\begin{cases} M(0, y) = 0, \\ M(x + 1, y) = M(x, y) + y = A(M(x, y), U_2^2(x, y)), \end{cases}$$

where A is the addition function of the preceding example.

3. **Exponentiation:** $E(x,y) = x^y$.
   With primitive recursion:

$$\begin{cases} E(x,0) = 1, \\ E(x,y + 1) = E(x,y) \cdot x = M(E(x,y),U_1^2(x,y)), \end{cases}$$

where M is the multiplication function of the preceding example.

The reader may wish to attempt defining by primitive recursion such functions as the factorial and the absolute difference.

In the following definition, the bracketed clauses are not part of the definition, but appear for the reader's convenience only.

17.8 Definition: The class of **primitive recursive functions** is defined recursively as follows:

(i) (*Basis*): The following basic recursive functions are primitive recursive: the *successor function* $[s(x) = x + 1]$, the *zero function* $[z(x) = 0]$, and the *selection functions* $[U_i^n (x_1, \ldots , x_n) = x_i$, for each $n \in \mathbb{P}$ and for each $i \in \{1, \ldots , n\}]$.

(ii) (*Generating Rule*): Any function obtained by *composition* or by *primitive recursion* on primitive recursive functions is itself primitive recursive.

(iii) Only functions definable by (i) and (ii), with a finite number of applications of (ii), are primitive recursive. ∎

It is quite easy to construct a program to compute each of the basic (primitive) recursive functions. The *successor function* concerns just incrementation by 1, the *zero function* is effected by the assignment of the constant value 0, and each of the *selection functions* assigns as the image whatever value appears in the specified position as an argument.

The mechanisms of *composition* and *primitive recursion* can be imitated by programs, using the corresponding definitions as blueprints.

Thus, every primitive recursive function is **computable;** i.e., it may be calculated in a finite chain of applications of the programs just mentioned, using programs as subroutines whenever necessary.

The class of primitive recursive functions is very extensive; perhaps surprisingly so when the rather simple definition of the class (Definition 17.8) is considered. Yet, the class of primitive recursive functions does not include all the *computable* functions, as is shown in the next section.

*17.10 RECURSIVE FUNCTIONS AND COMPUTABLE FUNCTIONS

The function defined below, known as **Ackermann's function,** is not primitive recursive, yet it is obviously "computable."

17.9 Definition: **Ackermann's function** f: $\mathbb{N}^2 \rightarrow \mathbb{N}$ is defined by

$$\begin{cases} f(0,n) = n + 1, \\ f(k,0) = f(k - 1,1), \\ f(k + 1, n + 1) = f(k, f(k + 1, n)). \end{cases}$$

The reader may wish to verify that $f(1,n) = n + 2$, for each n; $f(2,2) = 7$, and $f(3,3) = 61$. The reader is advised *not to try to evaluate* $f(4,4)$, unless the reader wishes to experience at first hand the meaning of "computable but not worth trying to compute."

A function whose domain is a subset of $\mathbb{N}$ is called a **partial function** on $\mathbb{N}$. Such a function is just *not defined* for the arguments not in its domain, if there are any. A function defined for each member of $\mathbb{N}$ is called a **total function** on $\mathbb{N}$.

We are about to introduce an operation on functions called **minimalization.** It is illustrated by the following.

Let $f(y,x) = |y^2 - x|$. Thus, for example, $f(1,1) = 0$, $f(1,2) = 1$; $f(2,2) = 2$; $f(2,4) = 0$.

Now, the question we ask is this: For a fixed value, say $x = 4$, what is the smallest value of $y \in \mathbb{N}$ so that $f(y,4) = 0$? The answer is clearly $y = 2$. Similarly, if $x = 1$, the smallest value of $y \in \mathbb{N}$ so that $f(y,1) = 0$ is $y = 1$.

We could recast this in a new function m (m for minimalization) whose value $m(x)$ at x is the smallest y such that $f(y,x) = 0$. Thus,

$$m(1) = 1, m(4) = 2, m(9) = 3, \text{ etc.}$$

Clearly, m is defined only for the perfect squares 1, 4, 9 etc. It is a *partial* function on $\mathbb{N}$.

---

*This section requires somewhat greater mathematical maturity.

Keeping the illustration in mind, the operation of **minimalization** takes as input a total function $f\colon \mathbb{N}^{n+1} \to \mathbb{N}$, whose value is written as

$$f(y,x_1,x_2, \ldots, x_n),$$

and produces from it a new function $m\colon \mathbb{N}^n \to \mathbb{N}$, whose value is written as

$$m(x_1,x_2, \ldots, x_n).$$

The values of m are found by the relation

$$m(x_1, \ldots, x_n) = \min_{y \in \mathbb{N}} \{y\colon f(y,x_1, \ldots, x_n) = 0\}.$$

The process producing the *partial function* m from the total function f is called **minimalization;** m itself is sometimes called the **minimalizing function of** f.

> *17.10 Definition:*   Extend the class of primitive recursive functions by allowing the operation of minimalization. The resulting class is that of the **partial recursive functions.**

This class contains not only functions which are not total, but also total functions which are not primitive recursive. An example of such a function is Ackermann's function

> *17.11 Definition:*   A function is **recursive** (also **general recursive**) iff it is a total function which belongs to the class of partial recursive functions.

As was the case with the operations of composition and primitive recursion, the operation of *minimalization* can also be imitated by a program without difficulty. In consequence, every recursive function can also be computed mechanically; it is said to be **effectively computable** (although the term is somewhat vague and largely intuitive in most treatments).

An important question to the computer scientist as well as to the logician is whether there is a function of which we conceive as effectively computable, but which is not recursive. The intuitive nature of the term effectively computable, in fact of what we would wish to mean by "effectively computable," renders a formal answer impossible. The following expression of sentiment is offered as an answer. It is often referred to as the **Church-Turing thesis,** as well as **Church's thesis.** It asserts that the answer to the last question is "no."

> *17.12 Church's Thesis:*   The class (intuitively thought of as that) of effectively computable functions is precisely the class of recursive functions.

List for
Further Reading

1. PURPOSE

a. Proofs As a rule, formal proofs of theorems are not presented in this book. Instead, intuitive arguments which are intended to enhance understanding are given to many results. The exceptions to this rule are occasional proofs presented for specific reasons, such as illustrating a type of formal argument, a proof with a particular flavor, or the study of proving theorems.

Thus, the duty this book assumes along formal lines ends with the presentation and organization of formal statements: definitions, theorems, algorithms.

The reader who finds a need or desire for access to formal proofs may then follow the references cited in this list. In most cases, there appear references both to elementary treatments and to "mature" ones, and the remarks accompanying the citation indicate the level(s) of discourse.

b. Additional Information In a book such as this *Companion* it is neither desirable nor feasible to treat a topic far beyond its utility to the vast majority of the prospective readers. Yet, the reader who has use for further information on the topic should be provided with sources for such information.

c. Advanced Treatment A minority of readers may find use for a treatment of a topic which is beyond the scope presented here, both in level and in extent. Sources for such treatment are identified in this list by the accompanying remarks.

2. CONTENT

This *List for Further Reading* is not intended as an exhaustive bibliography. In fact, too many references may be too overwhelming to the reader. Thus, the noninclusion of a source should not be regarded as a slight.

Only published books are included in this *List* and, it is hoped, ones which are easily accessible to the reader. Each book was selected both for its accessibility and for the perception that it contributes an ingredient or a flavor not present in the others.

In some cases, the result is a large contingent of citations and in others a more modest one.

Almost all the books cited in this *List* present formal proofs. The few which do not are so identified. They are included in this list for the excellence of other features they possess.

3. ORGANIZATION

This *List for Further Reading* is divided into self-contained segments, one for each chapter. Thus, a book may be included a number of times if it is useful in more than one chapter or in more than one topic. This repetition is intended so that the reader who is interested in the contents of one chapter (or of a portion of it) need not search through references not related to the chapter.

a. Citations for an Entire Chapter Such books contain information on a large number of the sections and topics in the chapter. A comment which appears with the citation [in square brackets] is intended to assess for the reader both the level of exposition and any particularly useful features of the book.

b. Citations for a Section These are of two kinds. If the book was also cited for the entire chapter, then there is a particular merit in the treatment of the subject(s) in this section. On the other hand, if the book was not cited for the entire chapter, then either the chapter-citations do not treat the subject(s) of the section adequately or there is particular merit in the treatment by this book.

c. A Citation for a Topic Inside a Section The topic appears in bold-face type (under the section number) and is easy to identify. Such a citation is often specific to the topic and not to the entire section.

When such a reference also appears as a citation for the entire chapter, then either some of the other references do not treat the topic at all or this book does so particularly well.

LIST OF BOOKS

1. Abian, A., *The Theory of Sets and Transfinite Arithmetic*. Philadelphia: Saunders, 1965. [Advanced. Detailed treatment of arithmetic of cardinals. Also a good introduction to sets and logic.]

2. Abbott, J. C., *Sets, Lattices, and Boolean Algebras*. Boston: Allyn and Bacon, 1969. [Clear advanced treatment of many topics; both elementary and advanced treatment of Boolean algebras.]

3. Arnold, B. H., *Logic and Boolean Algebra*. Englewood Cliffs, NJ: Prentice-Hall, 1962. [Intermediate and advanced. Applications to electrical networks, computer design, logic.]

4. Behzad, M., G. Chartrand, and L. Lesniak-Foster, *Graphs and Digraphs*. Boston: Prindle, Weber & Schmidt, 1979.

5. Berge, C., *Graphs and Hypergraphs*. New York: North-Holland, 1973. [Extensive.]

6. Berge, C., *Principles of Combinatorics*. New York: Academic Press, 1971.

7. Berge, C., *Theory of Graphs and Its Applications*. New York: Wiley, 1962.

8. Berman, G., and K. D. Fryer, *Introduction to Combinatorics*. New York: Academic Press, 1972. [Fairly elementary level; good for additional information but short on proofs.]

9. Birkhoff, G., *Lattice Theory (3rd ed.)*. New York: American Mathematical Society, 1967. [Detailed and concise. Intermediate and advanced.]

10. Birkhoff, G., and T. C. Bartee, *Modern Applied Algebra*. New York: McGraw-Hill, 1970.

11. Birkhoff, G. B., and S. MacLane, *A Survey of Modern Algebra (3rd ed.)*. New York: Macmillan, 1969. [Good introductory treatment.]

12. Brualdi, R. A., *Introductory Combinatorics*. New York: North-Holland, 1977. [Good elementary account, with applications.]

13. Carré, B., *Graphs and Networks*. Oxford: Clarendon Press, 1979. [Relatively elementary treatment of graphs (without proofs), with much information.]

14. Church, A., *Introduction to Mathematical Logic*. Princeton: Princeton University Press, 1956. [Thorough detailed formal treatment; intermediate and advanced.]

15. Clarke, B., and R. L. Disney, *Probability and Random Processes for Engineers and Scientists*. New York: Wiley, 1970. [Good, clear exposition with details, examples and proofs. Accessible.]

16. Clifford, A. H., and G. B. Preston, *The Algebraic Theory of Semigroups, Volume I*. Providence: American Mathematical Society, 1967. [Definitive algebraic treatment; intermediate and advanced. Includes a clear account of Light's test for associativity.]

17. Davis, P. J., *The Mathematics of Matrices*. Waltham, Mass.: Blaisdell, 1965. [Elementary, with more than sufficient detail.]

18. Denning, P., J. B. Dennis, and J. E. Qualitz, *Machines, Languages and Computation*. Englewood-Cliffs, NJ: Prentice-Hall, 1978. [Good detailed exposition. Elementary plus.]

19. Dieudonné, J., *Foundations of Modern Analysis*. New York: Academic Press, 1960. [Terse but good summary of a large number of topics.]

20. Even, S., *Algorithmic Combinatorics*. New York: Macmillan, 1973.

21. Feferman, S., *The Number Systems*. Reading, Mass.: Addison-Wesley, 1964. [Good illustrations and summaries. Clear and readable.]

22. Feller, W., *An Introduction to Probability Theory and Its Applications (3rd ed.)*. New York: Wiley 1968. [A classic. Complete. All levels— elementary to almost advanced.]

23. Flegg, H. G., *Boolean Algebra and Its Applications*. New York: Wiley, 1964. [Complete account; very readable; no proofs. Extensive chronological references up to 1962.]

24. Gelbaum, B. R., and J. M. H. Olmsted, *Counterexamples in Analysis*. San Francisco: Holden-Day, 1964.

25. Gill, A., *Applied Algebra for the Computer Sciences*. Englewood Cliffs: Prentice-Hall, 1976. [Good exposition, accessible. Proofs and additional information.]

26. Hall, M. Jr., *Combinatorial Theory*. Waltham, Mass.: Blaisdell, 1967. [Somewhat mature.]

27. Halmos, P. R., *Lectures on Boolean Algebras*. Princeton: Van Nostrand, 1963. [Very good advanced algebraic treatment.]

28. Halmos, P. R., *Naive Set Theory*. Princeton: Van Nostrand, 1964. [Advanced, clear and thorough; basic for every mathematician.]

29. Hamilton, N. T., and J. Landin, *Set Theory*. Boston: Allyn and Bacon, 1961. [Excellent account of elementary set theory. Good exposition of other topics.]

30. Hammer, P., and S. Rudeanu, *Boolean Methods in Operations Research and Related Areas*. New York: Springer-Verlag, 1968.

31. Harary, F., *Graph Theory*. Reading, Mass.: Addison-Wesley, 1969. [Detailed. Extensive author bibliography.]

32. Harary, F. (editor), *Proof Techniques in Graph Theory*. New York: Academic Press, 1969. [A collection of papers. Mature treatment. Excellent and extensive key-word indexed bibliography of graph theory up to 1969.]

33. Harrison, M. A., *Introduction to Formal Language Theory*. Reading, Mass.: Addison-Wesley, 1978. [Terse and mature treatment; clear and precise.]

34. Harrison, M. A., *Introduction to Switching and Automata Theory*. New York: McGraw-Hill, 1965. [Excellent and complete exposition; concise. Elementary and advanced.]

35. Hartmanis, J., and R. E. Stearns, *Algebraic Structure Theory of Sequential Machines*. Englewood Cliffs, NJ: Prentice-Hall, 1966.

36. Herstein, I. N., *Topics in Algebra*. Waltham, Mass.: Blaisdell, 1964. [Good elementary and intermediate treatment.]

37. Hohn, F. E., *Applied Boolean Algebra*. New York: Macmillan, 1966. [Elementary and intermediate. Detailed, with applications to logic, electrical networks, set algebra.]

38. Hohn, F. E., *Elementary Matrix Algebra (2nd ed.)*. New York: Macmillan, 1964.

39. Hopcroft, J. E., and J. D. Ullman, *Introduction to Automata Theory, Languages and Computation*. Reading, Mass.: Addison-Wesley, 1979. [Intermediate and advanced. Clear and precise.]

40. Jones, N. J., *Computability Theory: An Introduction*. New York: Academic Press, 1973. [Intermediate, terse and good.]

41. Kain, R. Y., *Automata Theory: Machines and Languages*. New York: McGraw-Hill, 1972. [Somewhat detailed, elementary treatment.]

42. Kelley, J. L., *General Topology*. Princeton: Van Nostrand, 1955. [In Chapter 0, an excellent terse but clear summary of terms, concepts and definitions. A classic.]

43. Kleene, S. C., *Mathematical Logic*. New York: Wiley, 1967. [Sweeping. Technical and advanced. Clear exposition.]

44. Kohavi, Z., *Switching and Finite Automata Theory (2nd ed.)*. New York: McGraw-Hill, 1978. [Elementary clear exposition without proofs. Leans to the engineering approach.

45. Kurosh, A. G., *The Theory of Groups*. New York: Chelsea, 1955 and 1960. [Mature treatment and detailed.]

46. Liu, C. L., *Elements of Discrete Mathematics*. New York: McGraw-Hill, 1977. [Good detail, with proofs. Not too advanced.]

47. Liu, C. L., *Introduction to Combinatorial Mathematics*. New York: McGraw-Hill, 1968. [Details. Elementary and intermediate levels.]

48. London, K. R., *Decision Tables*. New York: Auerbach, 1972.

49. MacMahon, P. A., *Combinatory Analysis, Volume I*. Cambridge: Cambridge University Press, 1915. [A classic. Mature treatment. Detailed treatment of distribution, calculus of binomial coefficient, permutations.]

50. Manna, Z., *Mathematical Theory of Computation*. New York: McGraw-Hill, 1974. [Definitive. Intermediate to mature treatment.]

51. McCoy, N. H., *Fundamentals of Abstract Algebra*. Boston: Allyn and Bacon, 1972. [Readable and detailed.]

52. Munroe, M. E., *Introduction to Measure and Integration (2nd ed.)*. Reading, Mass.: Addison-Wesley, 1971. [Advanced.]

53. Olmsted, J. M. H., *The Real Number System*. New York: Appleton-Century-Crofts, 1962. [Excellent and innovative account; clear, concise and complete.]

54. Preparata, F. P., and R. T. Yeh, *Introduction to Discrete Structures*. Reading, Mass.: Addison-Wesley, 1973. [Fairly elementary; accessible to the average reader.]

55. Quine, W. V. O., *Set Theory and Its Logic*. Cambridge, Mass.: Belknap, 1969. [Mature and clear treatment of many useful topics.]

56. Riordan, J., *An Introduction to Combinatorial Analysis*. New York: Wiley, 1958. [Clear, mature and thorough treatment, with proofs. Enumeration of rooted trees.]

57. Rogers, H. Jr., *Theory of Recursive Functions and Effective Computability*. New York: McGraw-Hill, 1967. [Mature treatment, but intuitively clear.]

58. Rudeanu, S., *Boolean Funtions and Equations*. New York: North-Holland, 1974. [Intermediate and advanced. A full treatment.]

59. Ryser, H. J., *Combinatorial Mathematics (Carus Monograph Number Fourteen)*. New York: Wiley, 1963. [Mature treatment. Good account of Boolean matrices.]

60. Sahni, S., *Concepts in Discrete Mathematics*. Camelot Lane, NE: Camelot, 1981.

61. Sikorski, R., *Boolean Algebras (3rd ed.)*. New York: Springer-Verlag, 1969. [Definitive; advanced.]

62. Stewart, I., and D. Tall, *The Foundations of Mathematics*. Oxford: Oxford University Press, 1977. [A clear intermediate to mature treatment.]

63. Stoll, R. R., *Set Theory and Logic*. San Francisco: Freeman, 1963. [Somewhat mature but easy to follow. Good exposition.]

64. Stoll, R. R., *Sets, Logic and Axiomatic Theories*. San Francisco: Freeman, 1961. [Not too advanced; good exposition. Particularly noteworthy is the treatment of Boolean algebra—detailed, from fundamentals to advanced.]

65. Suppes, P., *Axiomatic Set Theory*. Princeton: Van Nostrand, 1960. [Mature treatment of high expository quality.]

66. Suppes, P., *Introduction to Logic*. Princeton: Van Nostrand, 1964. [An excellent introductory book on logic and intuitive set theory.]

67. Suppes, P., and S. Hall, *First Course in Mathematical Logic*. New York: Blaisdell, 1964. [Elementary introduction and more intermediate treatment on inference, quantification, truth tables; considerable material on logic.]

68. Swamy, M. N. S., and K. Thulasiraman, *Graphs, Networks and Algorithms*. New York: Wiley, 1981. [Thorough, detailed treatment of graphs.]

69. Waerden, B. L. v.d., *Modern Algebra, Volume I*. New York: Ungar 1949. [A terse, mature treatment.]

70. Whitesitt, J. E., *Boolean Algebra and Its Applications*. Reading, Mass.: Addison–Wesley, 1961. [Good exposition on symbolic logic, truth sets, proofs (valid arguments), algebra of sets, switching algebra, circuits. Binary number system.]

71. Williams, G. E., *Boolean Algebra with Computer Applications*. New York: McGraw-Hill, 1970. [No proofs, but clear detailed elementary exposition. Number bases, binary arithmetic and other bases, laws of Boolean algebra, truth tables, simplification of Boolean functions.]

72. Zehna, P., and R. L. Johnson, *Elements of Set Theory*. Boston: Allyn and Bacon, 1962. [Elementary and intermediate; good account with details, proofs and additional information.]

LIST OF CITATIONS

*(Numbers refer to the **List of Books,** above. The mark $\supset$ denotes a subtopic of a section.)*

Index